Industrial Applications of Soft Computing

Studies in Fuzziness and Soft Computing

Editor-in-chief
Prof. Janusz Kacprzyk
Systems Research Institute
Polish Academy of Sciences
ul. Newelska 6
01-447 Warsaw, Poland
E-mail: kacprzyk@ibspan.waw.pl
http://www.springer.de/cgi-bin/search_book.pl?series=2941

Further volumes of this series can be found at our homepage.

Vol. 49. L.I. Kuncheva
Fuzzy Classifier Design, 2000
ISBN 3-7908-1298-6

Vol. 50. F. Crestani and G. Pasi (Eds.)
Soft Computing in Information Retrieval, 2000
ISBN 3-7908-1299-4

Vol. 51. J. Fodor, B. De Baets and P. Perny (Eds.)
Preferences and Decisions under Incomplete Knowledge, 2000
ISBN 3-7908-1303-6

Vol. 52. E.E. Kerre and M. Nachtegael (Eds.)
Fuzzy Techniques in Image Processing, 2000
ISBN 3-7908-1304-4

Vol. 53. G. Bordogna and G. Pasi (Eds.)
Recent Issues on Fuzzy Databases, 2000
ISBN 3-7908-1319-2

Vol. 54. P. Sinčák and J. Vaščák (Eds.)
Quo Vadis Computational Intelligence?, 2000
ISBN 3-7908-1324-9

Vol. 55. J.N. Mordeson, D.S. Malik and S.-C. Cheng
Fuzzy Mathematics in Medicine, 2000
ISBN 3-7908-1325-7

Vol. 56. L. Polkowski, S. Tsumoto and T.Y. Lin (Eds.)
Rough Set Methods and Applications, 2000
ISBN 3-7908-1328-1

Vol. 57. V. Novák and I. Perfilieva (Eds.)
Discovering the World with Fuzzy Logic, 2001
ISBN 3-7908-1330-3

Vol. 58. D.S. Malik and J.N. Mordeson
Fuzzy Discrete Structures, 2000
ISBN 3-7908-1335-4

Vol. 59. T. Furuhashi, S. Tano and H.-A. Jacobsen (Eds.)
Deep Fusion of Computational and Symbolic Processing, 2001
ISBN 3-7908-1339-7

Vol. 60. K.J. Cios (Ed.)
Medical Data Mining and Knowledge Discovery, 2001
ISBN 3-7908-1340-0

Vol. 61. D. Driankov and A. Saffiotti (Eds.)
Fuzzy Logic Techniques for Autonomous Vehicle Navigation, 2001
ISBN 3-7908-1341-9

Vol. 62. N. Baba, L.C. Jain (Eds.)
Computational Intelligence in Games, 2001
ISBN 3-7908-1348-6

Vol. 63. O. Castillo, P. Melin
Soft Computing for Control of Non-Linear Dynamical Systems, 2001
ISBN 3-7908-1349-4

Vol. 64. I. Nishizaki, M. Sakawa
Fuzzy and Multiobjective Games for Conflict Resolution, 2001
ISBN 3-7908-1341-9

Vol. 65. E. Orłowska, A. Szalas (Eds.)
Relational Methods for Computer Science Applications, 2001
ISBN 3-7908-1365-6

Vol. 66. R.J. Howlett, L.C. Jain (Eds.)
Radial Basis Function Networks 1, 2001
ISBN 3-7908-1367-2

Vol. 67. R.J. Howlett and L.C. Jain (Eds.)
Radial Basis Function Networks 2, 2001
ISBN 3-7908-1368-0

Vol. 68. A. Kandel, M. Last and H. Bunke (Eds.)
Data Minining and Computational Intelligence, 2001
ISBN 3-7908-1371-0

Vol. 69. A. Piegat
Fuzzy Modeling and Control, 2001
ISBN 3-7908-1385-0

Vol. 70. W. Pedrycz (Ed.)
Granular Computing, 2001
ISBN 3-7908-1387-7

Kauko Leiviskä
Editor

Industrial Applications of Soft Computing

Paper, Mineral and Metal Processing Industries

With 118 Figures
and 27 Tables

Springer-Verlag Berlin Heidelberg GmbH

Prof. Kauko Leiviskä
University of Oulu
Department of Process Engineering
P.O. Box 4300
90014 Oulun yliopisto
Finland
kauko.leiviska@oulu.fi

ISSN 1434-9922

DOI 10.1007/978-3-7908-1822-2

Cataloging-in-Publication Data applied for
Die Deutsche Bibliothek – CIP-Einheitsaufnahme
Industrial applications of soft computing: paper, mineral and metal processing industries; with 27 tables / Kauko Leiviskä Ed. – Heidelberg; New York: Physica-Verl., 2001
(Studies in fuzziness and soft computing; Vol. 71)

Originally published by Physica-Verlag Heidelberg New York in 2001
MyCopy version of the original edition 2001

Hardcover Design: Erich Kirchner, Heidelberg
www.springer.com/mycopy

Preface

My first contact with Soft Computing dates back to the end of 1970's. I got a letter from a foreign colleague who asked, if I were interested to study a new control methodology, fuzzy logic control. I had just finished with my Licentiate of Technology thesis that had led to the first commercial control system for the lime kiln. I had also started my Doctor of Technology studies and selected production scheduling algorithms for the pulp mill as my topic. So, I had to turn his proposal down.

Next contact was about seven years later, 1985. I was working with the lime kiln again. This time, one my students was testing the expert system technology for the consulting application on the lime kiln. The program system we were using had a fuzzy system option, and we made a consulting fuzzy expert system for the lime kiln operation. The system was never installed on the actual kiln.

I got my own laboratory to manage in 1990 and we started working with the its strategy. My interest to knowledge-based system and to fuzzy logic and neural networks had grown during years. We had also discussions with companies and financing sources and, together with the interest, we chose Intelligent Methods, how we called them at this time, our strategic growth area.

After couple of years, we were working with the lime kiln control again. This time we used fuzzy logic control and the studies led to our first successful industrial application on this new area. So, it is no wonder that also one of the chapters in this book deals with the use of Soft Computing methods in lime kiln.

This book concentrates on the application of Soft Computing methods in Finland. Finland experienced a boom of neural network and fuzzy technologies during 1990's. There were several reasons for this. Finnish companies have been pioneering in many areas of industrial control applications for decades. Especially this agrees with mining and metals and paper industries. Companies have seen the benefits of better control and improved information, and also in this case it was easy to find industrial partners ready for testing the new technology. Also the research policy that the State of Finland was carrying out was positive for new technologies. As a concrete example, the National Technology Agency, Tekes, launched an extensive research and technology development project concerning intelligent and adaptive systems. Its effect for the growth of fuzzy logic and neural network applications in the Finnish industry was a tremendous one.

While choosing papers for this book I was in front of multitude of possibilities. I have tried to balance the material at least in two ways. First, I tried to find new interesting industrial applications that have not been too reported earlier in the literature. Secondly, I tried to give a chance to as many research groups as possible to tell about their research.

After four short introductory chapters, applications in mineral and metal processing start. The first application tells about fuzzy logic control of the sintering plant. It is a good example, how a commercial software is used controlling a complicated process. The second example tells about an application Kohonen self-organising map to describe the behaviour of wall temperature profiles in the blast furnace. It shows the strength of SOM in visualising complicated process phenomena in giving improved information for operators. The third example tells about the smelt shop control. It is an example, how neural network models are integrated with existing models to a hybrid system with increased performance. It shows the ability of neural networks to cope with huge amounts of data.

The fourth example of this topic takes us to mineral processing. It shows the results of a long research line concerning the modelling and control of a rotary dryer for minerals drying using conventional and advanced methods. The fifth, and the final example introduces even more difficult example from minerals processing. It deals with using image processing and an expert system in flotation plant control.

Next we turn to the paper industry. These applications start from modelling. The first paper in this sequence deals with the comparison of different fuzzy modelling methods using continuous pulp cooking as an example. The second example concerns with a conventional fuzzy logic control. It concern with a fuzzy control system for TMP-pulping installed at the Swedish mill. The third example is the "obligatory" lime kiln example. The existing control system combines fuzzy logic control, neural networks and an expert system framework. This system is in operation and the results have been good. The forth example is once again dealing with TMP- pulping. It shows how fuzzy logic can be combined with Statistical Process Control to develop a flexible system to help operators in process monitoring. Finally comes an example from the paper machine control. It concentrates on paper grade changes that are usually too complicated for manual operation and the risks of production losses are also high in automatic changes.

As the Editor of this book, I have many thanks to give. I thank the Series Editor for the possibility to write this book and good advises along the way. I thank my writer colleagues for good co-operation and patience, because the road was long.

I thank my wife, Eija and my daughter, Minja, for just being there. This effort took a lot of our time together. Especially, I thank my son, Marko, for collecting the background material for the Chapter "Applications in Paper Industries". It was a good work.

Oulu, November 20, 2000 Kauko Leiviskä

Contents

Part III: Paper Industries

Part I

Introduction

Basics of Soft Computing Methods

Kauko Leiviskä
University of Oulu, Control Engineering Laboratory, P.O.Box 4300, FIN-90014 Oulun yliopisto, Oulu, Finland

Zadeh (1994) introduced the term "Soft Computing" for the first time. He used the term to mean systems that "exploit the tolerance for imprecision, uncertainty, and partial truth to achieve tractability, robustness, low solution cost, and better rapport with reliability". It includes fuzzy logic, neural computing, evolutionary computing and probabilistic computing as main methodologies. Like any other concept, also Soft Computing has many definitions.

Fuzzy sets provide a unified framework for taking into account the gradual or flexible nature of variables, and the representation of incomplete information (Dubois and Prade 1992). Fuzzy logic control (FLC) has been one of the most active and fruitful areas for applications of fuzzy set theory. A wide variety of fuzzy development and simulation environments are available. Most of the systems are based on fuzzy rules, Mamdani- (Driankov et al., 1993) or Tagaki-Sugeno type (Takagi and Sugeno 1985). Applications vary: software sensors, alarming systems, process diagnostics, process control at different control levels, quality control, and production scheduling. Also application environments vary from microcontrollers to automation systems and extensive expert systems for quality control and scheduling.

Rules are generated in the same way as in conventional knowledge-based systems using expert knowledge. If no prior knowledge exists, the rules and membership functions can be extracted from the data (Babuska and Verbruggen 1995). Several technologies are available, for instance fuzzy c-means (FCM) clustering (deOliviera 1993). In rule generation, however, two different cases exist. The first is the data intensive case, where a lot of measurements covering the whole range of variables are utilised. This calls for good data mining methods and automatic rule generation. The opposite case is the poor data case. If the processes are slow, experimenting is expensive and data acquisition possibilities poor, the situation is easily met where very few data are available. This situation calls for methods that can combine the few data available with the expert knowledge.

Neural networks (NN) get inspiration from biological nervous systems and mathematical theories for learning (Huang and Chang 1995). They are characterised by their learning ability and a parallel distributed structure. Neural networks can also be considered as black box modelling methods.

Various methods have been developed for neural computing. Back propagation (Rummelhart et al. 1986) and self-organising maps (Kohonen 1997) are the best known methods. Learning algorithms have a close relation to optimisation methods, e.g. back propagation method uses a gradient descent. A huge number of programming and development tools is available both commercially and in public domain.

Neural networks are often combined with other approaches, for instance the ANFIS method (Adaptive-Network-based Fuzzy Inference Systems) which is suitable for tuning of membership functions (Jang 1993). Various other neural learning methods (deOliviera 1993; Glorennec 1994; Jang 1992), orthogonal least squares (Wang 1994), inductive learning (Ross 1995) or knowledge-based networks (Lacher et al. 1992) exist.

Neural networks are most useful in data intensive cases: in modelling, classification and feature extraction, They suffer from poor data. A lot of control applications also exist (Hunt et al. 1992).

Genetic algorithms (GA) are optimisation methods, which lend their search method and vocabulary from biology (chromosomes, genes, selection, crossover, and mutation) (Goldberg 1989). GA maintains and develops a population of potential solutions, which are called individuals (or genotypes, or structures). The algorithm (selection, crossover, mutation) is a reasonable and efficient way for processing population of alternatives.

Genetic algorithms are also often combined with other methods of soft computing. For expert systems, optimising the rule base is a possible application. They are also used in tuning fuzzy systems. Network structures of neural networks could be another application area, but the execution will be always very slow since both GA and NN require a lot of computation.

This book considers also "classical" expert systems, because a lot of applications exist in process industries. In rule-based expert systems, the domain knowledge is represented as sets of rules that are checked against a collection of facts or knowledge about the current situation (Waterman 1986). They rely on expert's knowledge and inference mechanisms that in a way simulate the human reasoning. The method suffers from two disadvantages: the knowledge acquisition principle, which may lead to really massive rule bases and weaknesses in the uncertainty processing which is needed in real world problems. Tables 1 and 2 show some properties of Soft Computing methods. Following chapters concern with the basics of fuzzy logic, neural networks and genetic algorithms.

1
Fuzzy Logic Control

Conventional dual logic does not accept that some item may partially belong to two or more separate sets. It starts from the idea of "crispness", each item has its place in one set, only. The real world, however, behaves in another way; situations are very often vague and uncertain, the further state of the system might not be known for certain, etc. Our natural languages have vague meanings for words. The world is not black and white, but it has several shades of grey.

Table 1. Properties of Soft Computing methods.

	ES	FL	NN	GA
Advantages	Extracting knowledge Prototyping Explanations	Handling of uncertainty Natural compromises Easy to built small systems Explanations	Automatic modelling Black box modelling Precision	Large search space Design and optimisation capabilities
Weaknesses	Complexity Handling of uncertainty Testing	Tuning of complex systems Doubts on stability	Apply to fragments No explanation Safety	Computer requirements No good for control
Design Information	Rules	Rules Data	Data	Populations Fitness function

Table 2. Connections between methods and hybridisation possibilities. Table should be read by rows; what the technology has to offer to others.

	ES	FL	NN	GA
ES		Rules	Knowledge-based networks	Knowledge-based systems
FL	Inference Uncertainty handling		Hybrid controller structures	
NN		Neural-fuzzy learning		
GA	Optimising rule base	Optimising rule base Tuning membership functions	Optimising NN structures	

Fuzzy logic is not a new approach. Zadeh first introduced the theory of fuzzy sets in 1965 (Zadeh 1965). The introduction gave a strict mathematical framework for studying vague concepts and phenomena. Fuzzy logic control is a component of artificial intelligence. First control applications developed primarily during the 1970's (Mamdani 1974). Many industrial applications have originated in Japan.

Commercial products based on fuzzy sets theory are also available. These products include microprocessors with special designs for fuzzy reasoning. Applications vary from different artificial intelligence applications to sensors and home electronics. Commercial industrial controllers for fuzzy logic control are also available today. Various computer programs and program packages for the development and installation of fuzzy logic control exist. Universities and research institutes are conducting considerable research work. The number of actual industrial applications is growing.

1.1 Concept of Fuzziness?

Fuzziness is a concept that tries to describe inaccuracies and uncertainties that are common in daily discussions. It represents a set of methods that makes the use of linguistic variables and descriptors in process control possible. For example, concepts such as young, middle-aged, and old are common when speaking about someone's age. "Age" can be considered as a linguistic variable, when its values are given as words and sentences rather than in a numerical form (Zadeh 1975) and they do not have any exact value but rather a range of values. Linguistic values, young, middle-aged and old, before are sometimes called (linguistic) fuzzy labels and they correspond to fuzzy sets. These fuzzy sets are described by membership functions. Fig. 1 shows a group of membership functions for the linguistic variable "temperature"; low, medium and high. Looking at the range of numerical values they cover, it seems that this example might consider, for instance, a Finnish sauna.

This picture means that no exact single-valued expression applies for instance to the fuzzy label "medium". The presentation can have some flexibility – a gradual change from one linguistic value to another. In the conventional theory of sets, an element either belongs or does not belong to some set. In the theory of fuzzy sets, there is no exact limit for an element to belong or not to belong to a fuzzy set. Instead, the degree of membership shows a "partial" inclusion into a fuzzy set that can vary from zero to one.

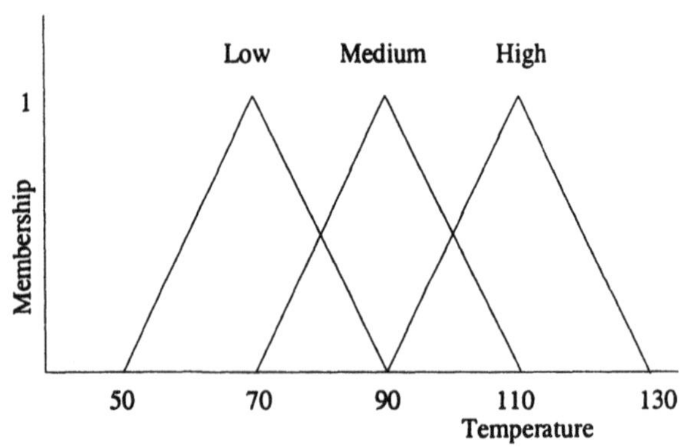

Fig. 1. Membership functions for three temperature labels.

The difference is obvious when comparing the fuzzy description in Fig. 1 with the crisp mathematical description in Fig. 2. One can interpret these figures so that the crisp valued mathematics is a subset of more versatile fuzzy mathematics. This is because common mathematics cannot handle expressions such as "almost medium" or "little over medium" that are possible in fuzzy logic.

Instead of triangular membership functions also other forms, e.g. trapezoidal functions of Fig. 3 can be used.

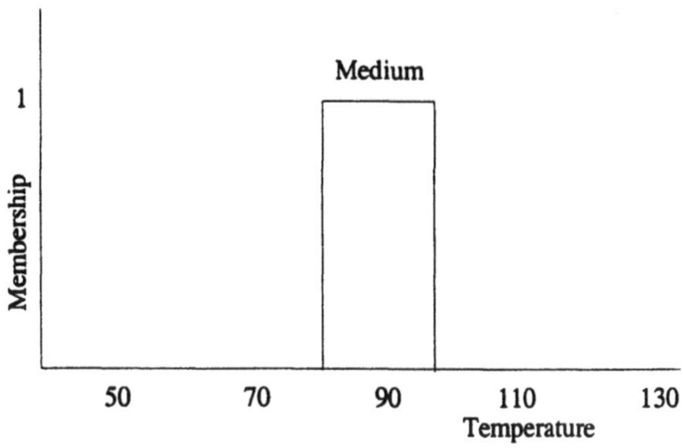

Fig. 2. Crisp description for the variable "medium".

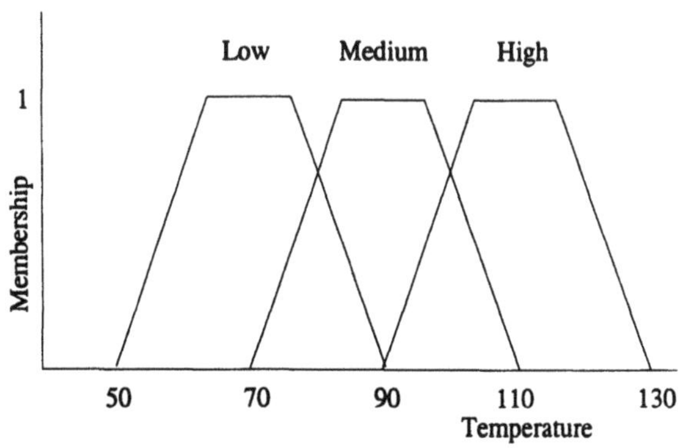

Fig. 3. Trapezoidal membership functions.

1.2 Fuzzy Logic Controller

Fuzzy logic control is a tool that can transfer an operator's knowledge in a linguistic form into a controller. This means that operator experience becomes a part of control algorithms in the form used in the human reasoning. Instead of exact numerical values, fuzzy logic controllers operate with values as low, high, medium, sufficient, etc. This form of presentation also helps collect knowledge from people who speak in qualitative terms rather than in numerical values.

According to Fig. 4, a fuzzy logic controller has three main parts: fuzzification, fuzzy reasoning, and defuzzification. In its operation, a fuzzy logic controller uses membership functions and rule sets.

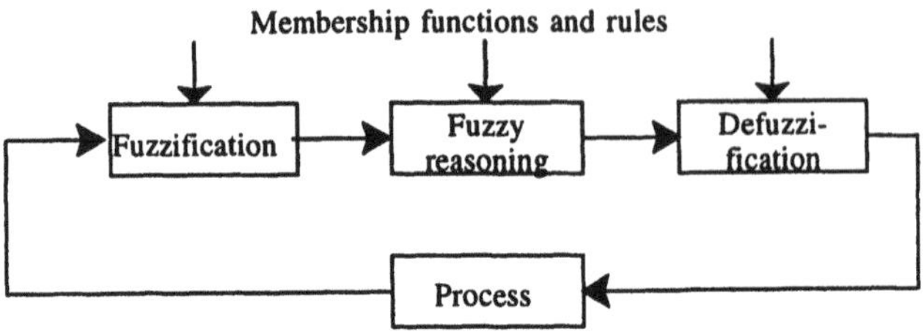

Fig. 4. Fuzzy logic controller.

1.2.1 Fuzzification

Fuzzification means the methods and principles used when converting exact measurement values to linguistic fuzzy variables. In principle, this happens by dividing the measurement range into as many subareas (labels) as seems reasonable. For instance, in the above case, the temperature has three fuzzy labels:

- Low, when it is bigger than 50°C and smaller than 90°C
- Medium, when it is bigger than 70°C and smaller than 110°C
- High, when it is bigger than 90°C and smaller than 130°C.

This selection is arbitrary and case dependent. In this "sauna example" it might even be dependent on personal preferences.

Fuzzification means connecting of actual process measurements with fuzzy labels and membership values. According to Fig. 5, temperature of 95°C in this case corresponds to

- "High" temperature with the membership of 0.25
- "Medium" temperature with the membership of 0.75.

1.2.2 Rule Base

Fuzzy reasoning uses rules for the fuzzified measurement values. Three main variants exist. Linguistic fuzzy rules are called Mamdani rules, when both the antecedent and the consequent parts are fuzzy propositions. Linguistic fuzzy if-then rule can be given in a general form:

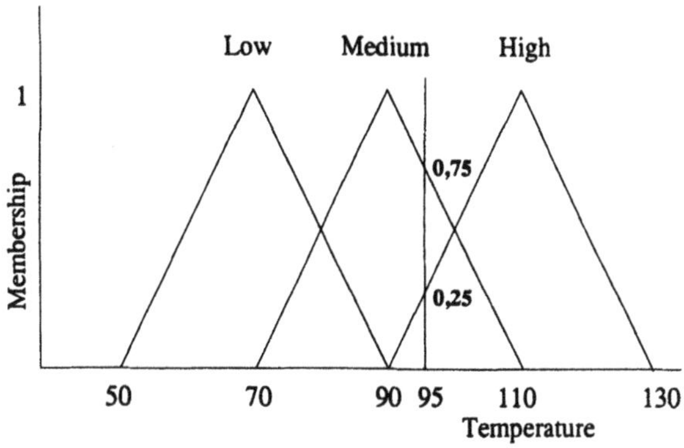

Fig.5. Example of fuzzification.

R_i: **If** x is A_i **then** y is B_i, $i=1,2,\ldots,K$, (1)

where x is the antecedent variable (input). Similarly, y is a consequent variable (output). A_i and B_i are linguistic terms (labels), which are defined by fuzzy sets $\mu_{Ai}(x)$:$X \rightarrow [0,1]$ and $\mu_{Bi}(y)$:$Y \rightarrow [0,1]$. R_i denotes the *i*th rule, and K is the amount of rules in the rule base.

In the previous case the antecedent propositions contain fuzzy sets, which are directly defined in the vector domain X. More often, the rules are in a decomposed form where the antecedent is defined as a combination of simple fuzzy propositions on the individual components x_i, i=1,2,..,p of the vector x. By using the logical operators (conjunction, disjunction and negation) a desired compound proposition can be constructed. For example:

R_i: **If** x_i is A_{i1} **or** x_2 is A_{i2} **and** x_3 is **not** A_{i3} **then** y is B_i. (2)

where the degree of fulfilment of the rule is calculated using the appropriate t-norm, t-co-norm and the complement operator. The necessary mathematics is found in any textbook (e.g. Driankov et al. 1993; Wang 1994).

In Singleton fuzzy rules the consequent fuzzy sets B_i of a Mamdani rule can be reduced to fuzzy singletons and represented as real numbers b_i:

R_i: **If** x is A_i **then** $y = b_i$ (3)

When Singleton fuzzy rule is compared with a Mamdani rule, the number of distinct singletons in the rule base is usually not limited. It means that each rule may have its own singleton consequent.

Third possibility is to use Takagi-Sugeno (TS) rules that are suitable for approximation of a large class of non-linear systems. In Takagi-Sugeno fuzzy rule, the rule consequents are crisp functions of the rule inputs:

R_i: **If** x is A_i **then** $y_i = f_i(x)$, (4)

where x is input and y is output variable. The consequent functions f_i are usually chosen as instances of a suitable parameterised function, whose structure remains equal in all the rules and only the parameters change. Affine linear form is a simple and practically useful parameterisation:

$$y_i = a^T x + b_i, \tag{5}$$

where a_i is a parameter vector and b_i is a scalar offset. The previous rule can be called as an affine TS rule. A special case occurs when offset $b_i = 0$. In such a case the rule can be called a homogenous TS rule. This kind of rule has not so good approximation capabilities than the affine TS rule. If the parameter vectors $a_i=0$, $i=1,\ldots,K$ the consequents are constant functions and the rule is the same as Singleton rule before.

The size of the rule base depends on the number of input variables and the number of their linguistic labels. If the number of input variables is three (x_1, x_2, x_3) and they all have three linguistic labels (low, medium, high), the complete rule base has 27 rules according to the following tables.

x_1 is "low"

x_2/x_3	low	medium	high
low	R1	R2	R3
medium	R4	R5	R6
high	R7	R8	R9

x_1 is "medium"

x_2/x_3	low	medium	high
low	R10	R11	R12
medium	R13	R14	R15
high	R16	R17	R18

x_1 is "high"

x_2/x_3	low	medium	high
low	R19	R20	R21
medium	R22	R23	R24
high	R25	R26	R27

Now, if the temperature in Fig. 5 corresponds to the variable x_1 in the previous rule base, 18 rules are triggered when the temperature is 95°C!

1.2.3 Defuzzification

Defuzzification is necessary because when several rules are triggered also several fuzzy sets corresponding to the output variable are "true" with different membership values. Fig. 6 shows a very simple case corresponding to the temperature control of previous figures. Here, only two fuzzy sets are supposed to describe the output variable (heat input, either heating or cooling): it is either kept constant or increased. In this case, the inference with the rule base has lead to two opposite requirements for the output:

- Keep it constant with the membership value of 0.50
- Increase it with the membership value of 0.75.

Clearly the actual output value is one kind of compromise between these values. This compromise is mostly calculated with some kind of averaging operator. Several methods are available for this (Driankov et al. 1993; Wang 1994):

- Centre of area
- Centre of sums
- Centre of largest area
- First of maxima
- Middle of maxima
- Height (weighted sum of peak values).

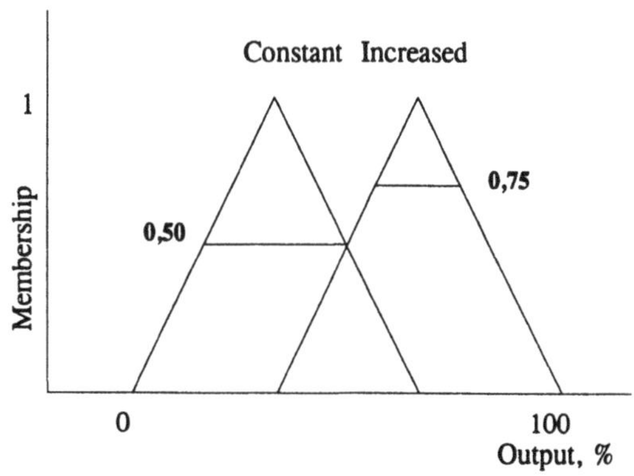

Fig. 6. Results of fuzzy inference before defuzzification.

1.3 Fuzzy Control Strategies

Originally, fuzzy logic control was introduced to mimic operator actions and this way to automate "the best operator practice" into the process. The processes that are very suitable for fuzzy logic control (see e.g. Verbruggen and Brujn 1997):

- Are only partly known
- Are difficult to model with conventional methods
- Have difficulties with measurements
- Are highly non-linear.

Also the production environments are changing. For instance, paper mills operate with higher and higher production and producing several paper grades with shortening time for changes. This calls for improved monitoring and control capabilities. In electronics production, short life-cycle products put more and more emphasis on fast ramp-up periods, once again meaning the necessity for good coordinating controls. Also the amount of information is increasing. Ten years ago, we suffered from too few data, but nowadays it is found in abundance. Three areas seem to gain importance:

- Quality and environment
- Measurements; fusion of measurements and software sensors
- Heuristic data.

All these areas provide also good applications for fuzzy logic. Looking at process control, several application areas are already found:

- Replacing low-level (e.g. PI-) controllers
- Taking care of the upper level supervisory control
- Taking care of exceptional cases, while conventional controllers are used in normal cases
- Tuning of conventional controllers
- Adaptive fuzzy control
- Model reference fuzzy control.

Some of these areas are included in the later chapters.

2 Neural Networks

2.1 Basic Definitions

A neural network consists of a big amount of simple parallel processing units connected with each other. Each connection between the units has its weight that describes the strength of the connection. The weights are tuned, i.e. the neural network is learning the certain application. Fig. 7 shows this kind of simple processing unit and its connections with the environment. In the figure, a_i describes the activity level of inputs and w_{ji} denote the connection weights.

Next figure (Fig. 8) shows a multi-layer network consisting of three layers. The lowest level is the input layer and the uppermost level is the output layer and the level between them is called the hidden layer. The hidden layer does not have outside connections, but it has importance in the network learning. Inputs are coming only to the input layer.

The number of hidden layers and the number of neurons on each layer have the greatest effect on the network performance. Also the chosen learning algorithm has its effect.

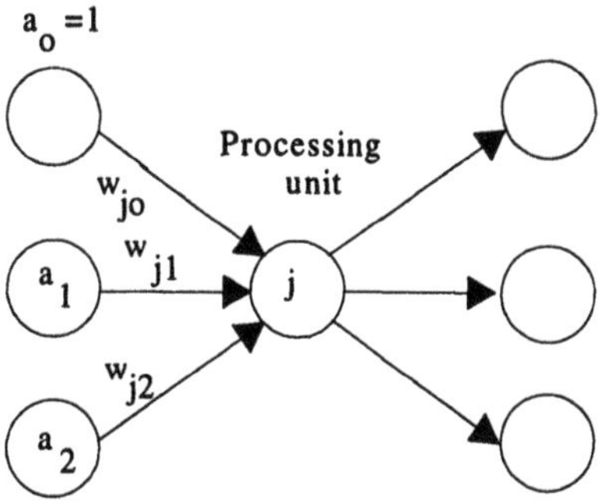

Fig. 7. Processing unit in a neural network.

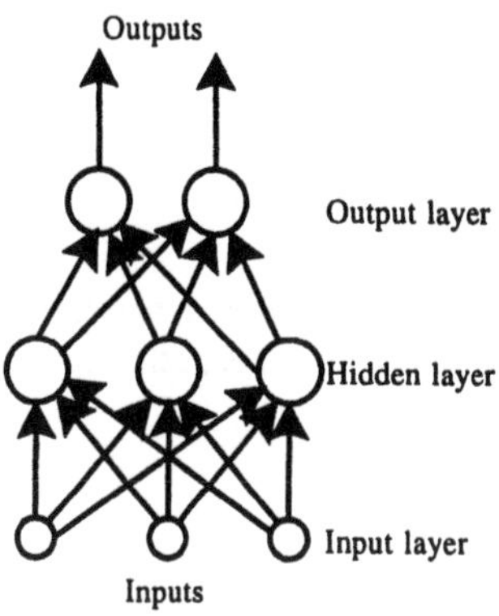

Fig. 8. A multi-layer network.

2.2 Principles of Learning

The short-term knowledge in neural networks is stored in the states of neurons and their connections correspond to the long-term knowledge. In this connection learning means either the adaptation of weights or adding or removing connections between neurons. This divides networks to static and dynamic neural networks. In static networks, the weights are changing, but the connections remain unchanged. In dynamic networks also the connections change.

Some networks learn from examples. A group of examples (the training set) is given to a network and the network tries to learn (generalise) them. The contents of training set vary from one application to another; it can consist of pictures, hand-written signals, speech, measurement data, etc. During learning optimum values for the weights in the neural network are determined. Learning is continued until the desired performance is achieved. In practice, this means that the network produces required outputs. The training set can be introduced to the network several times so the learning proceeds stage wise.

Learning algorithms can be classified into two main groups:

- Supervised learning
- Unsupervised learning.

In supervised learning the required outputs corresponding to the certain input set is known in advance. Initial weights are set randomly. In unsupervised learning no external reference signals (training set) are used.

2.3 Back Propagation Networks

Back propagation is the most usual network solution. Fig. 8 before showed an example of the back propagation network. It consists of at least two completely connected layers and often the number of layers exceeds two. In the previous figure the lowermost layer represents the input variables and it is the only layer connected to outside inputs. The uppermost layer is the output layer and the layer between them is the hidden layer:

Fig. 7 before presents one processing unit of the back propagation network together with its connections. Input units are at the left and output units at the right hand side. The activation level of the jth unit is denoted by a_j.

Training of the back propagation network occurs in two stages:

1. The forward propagation stage starts from the input values and calculates the activation levels of all neurons on the hidden layer utilising weighted sums and threshold functions. The output values are got from the output layer.

The calculation proceeds as follows: First, all the input connections to a processing unit are summed

$$S_j = \sum a_i w_{ji} \tag{6}$$

The activation level of the processing unit is calculated using the threshold function

$$a_j = f(S_j)\,. \tag{7}$$

The threshold function is usually a sigmoid (Fig. 9) that is relatively constant with small and big values, but varies sharply in the middle of the region

$$f(x) = \frac{1}{1+e^{-x}} \tag{8}$$

Other possibilities are (Wells 1995):

- Binary function
- Semi-linear function
- Radial basis function.

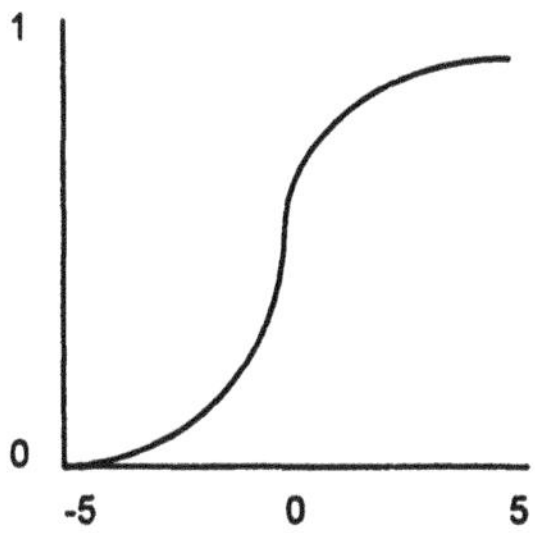

Fig. 9. Sigmoid function.

The activation level given by the above equations is set as output for each connection leaving the processing unit in question.

There is no summing operations on the input layer, but the activation levels of its units equal to input values. In some network a bias can be connected to any other layer except the output layer. The activation level of the bias equals to 1. This unit is connected to all the upper level units and it can improve the convergence of the network.

2. In the back propagation stage the weights of the network are corrected. Calculations start from the output layer and propagate through the whole network. For the output unit the error is calculated as follows

$$e_j = (t_j - a_j) f'(S_j), \tag{9}$$

where f' is the derivative of the sigmoid that puts the biggest corrections to the area where sigmoid is increasing with the highest speed. It means that the maximum of f' is achieved when x becomes zero.

On the hidden layer the error is calculated from the equation

$$e_j = \left[\sum_k e_k w_{kj}\right] f'(S_j). \tag{10}$$

Here e_k corresponds to the errors from the upper layer units that are connected to the hidden layer unit in question and w_{ki} is the corresponding weighting factor. Using the above calculated errors the changes in the next lower layer links (the corresponding weighting factors) are calculated. Training speed, C, is usually between 0.25 ... 0.75.

$$\Delta w_{ji} = C e_j a_i. \tag{11}$$

2.4 Kohonen Self-Organising Map

Kohonen self-organising map (SOM) algorithm is a well-known example of neural network models based on unsupervised learning. The network is trained without giving it any desired output. The SOM can be used in grouping and visualisation of input set.

The network of SOM consists of two layers: standard input layer and an output layer known as a "competitive Kohonen" layer. These two layers of neurons are usually organised into two-dimensional grids. Each neuron on the input layer is connected to every neuron on the competitive layer (Fig. 10).

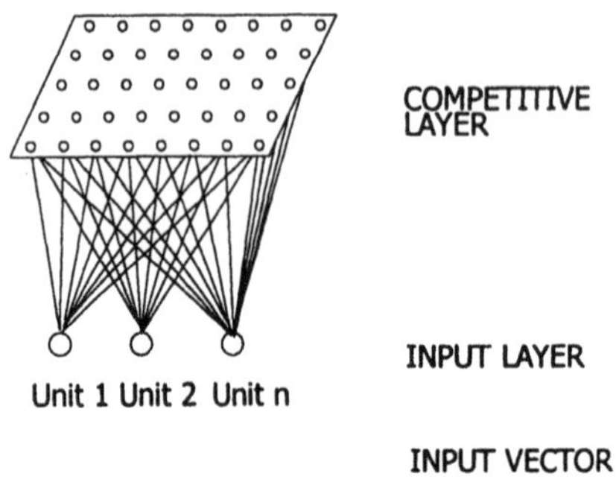

Fig. 10. Structure of SOM network with two-dimensional layers.

Each neuron has a neighbourhood, which has a significant meaning in training the network, as Fig. 11 shows.

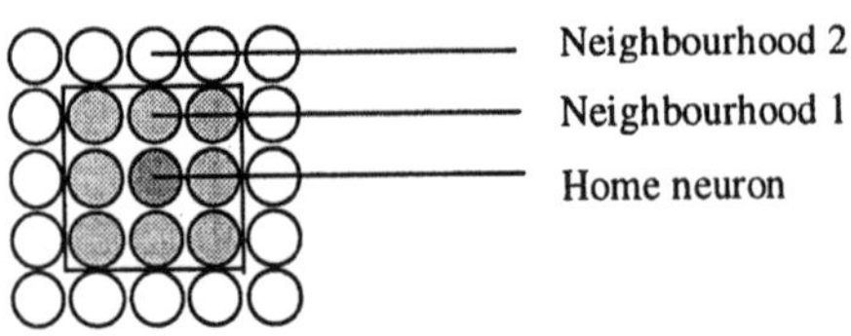

Fig. 11. Neuron and its neighbourhood.

Learning begins from a random starting point and the connection weight vectors are set to small random values. The learning rate is chosen and the neighbourhood is defined big enough to contain all the neurons. During training, both the learning rate and the size of the neighbourhood decrease. The winning node, which has the smallest Euclidean distance, is located among the competing neurons and its weight vectors and the neurons in its neighbourhood are updated. Those outside this neighbourhood are left unchanged. Iteration is continued until no further change in the output feature map is observable or until another termination condition is achieved. The resulting map shows the relationships between the input patterns given to the network.

2.5 Neural Control Strategies

Neural network controllers may consist of one or more neural networks The controller may also contain linear elements such as integrators or derivatives. It is worthwhile to use neural networks when the advantages compared to other methods can be clearly seen. Otherwise, it may be more reasonable to use some conventional method

Hunt (1992) gives a good survey of neural controllers and their structures. Following control structures have been used:

- Feedback network controller
- Optimal controller
- Predictive controller
- Internal model controller
- Model reference controller.

Conventional control implementation requires expensive and arduous detailed process analysis made beforehand. By using neural networks this can be at least partly avoided. Neural network controller can be applied also to processes, which have a complex mathematical model e.g. partial differential equations or no model at all.

In control applications neural networks offer some features that are of advantage:

- Fast parallel processing
- Analysis tool in addition to controller (no need for a mathematical model)
- Ease of implementation
- Good fault tolerance
- Extrapolation capability
- Robustness
- Adaptive systems by changing the network weights (but requires additional training).

Choosing the network structure for particular applications may be problematic. The choosing of the number of input, hidden and output neurons as well as the number of hidden layers and initial connection configurations and weights are determined through trial and error even though some empirical guidelines or rules of thumb are available. Care must also be taken not to overparametrise or overtrain the networks. The number of data points should always exceed by far the number of weights to be learnt. Overtraining is avoided by using separate training and testing data sets and by following the generalisation capability of the network. The quality of training data should also be considered carefully. A common statistical analysis is always recommended and the independence of input variables should always be taken care of.

3 Genetic Algorithms

3.1 Algorithm

Genetic algorithms (GA) have a biological basis. GA tries to imitate the nature's evolution mechanisms. The basic concepts in evolution are chromosomes, which delineate the structures of creatures, populations, natural selection – the survival of the fittest, mutation and crossing. The basic idea of this method is so called building block hypothesis: good parameter vectors include partial solutions, which can be connected to produce even better solutions.

Genetic algorithms (GA) are optimisation methods in which the stochastic search uses the structure based on biological principles referred to before (Ishigami et al. 1995; Linkens and Nyongesa, 1995; Michalewicz 1992]. GA maintains a population of potential solutions, which are called individuals (or genotypes, structures).

The GA typically starts by randomly generating its initial population of chromosomes. Each chromosome will be measured by its fitness value. On the basis of the fitness value, the chromosomes undergo genetic operations like selection, crossover and mutation. The crucial thing is that fitness function measures the right thing, i.e. it should be application dependent. The goal of genetic operations is to find a set of parameters that reach an optimal solution to the problem or to reach the constraints of the generations.

The genetic algorithm proceeds as follows:

- Create an initial population $P^0_{i,j}, i = 1,\dots,n, j = 1,\dots,m$ where n is the size of the population and m is the amount of bits in chromosomes.
- New chromosomes are produced by crossing individuals of population p^k .
- Generate mutations in population p^k.
- Define goodness of the population.
- Is the result good enough?
- If the result is satisfying, then stop.
- Else, choose the best individuals in population p^k_i for next cycle in other words, regeneration. Return to the step 2.

The algorithm goes on until the result is satisfactory or iterations reach the maximum.

3.2 Use of GA in Control

GA is not usually used alone in real-time control. One reason for this is that the method requires a lot of computation and thus consumes a lot of time. But, when using other methods to control the system, GA can be used to optimise their structures. E.g. GA can optimise the rule base of the expert system and the structure of the neural network. Zhou et al. (1997) shows an example of using GA in fuzzy modelling. According to Fig. 12 the system includes initialisation, parameter optimisation by genetic algorithm and fuzzy inference. First the initial values are defined: population size, maximum and minimum linguistic levels for the variables, working range, possibility of crossover, possibility of mutation and the iteration times in the initialisation stage. Then program generates the initial population, membership functions for the input and output variables and the coefficients for the rule base. Finally, the application starts and continues until the termination condition is met.

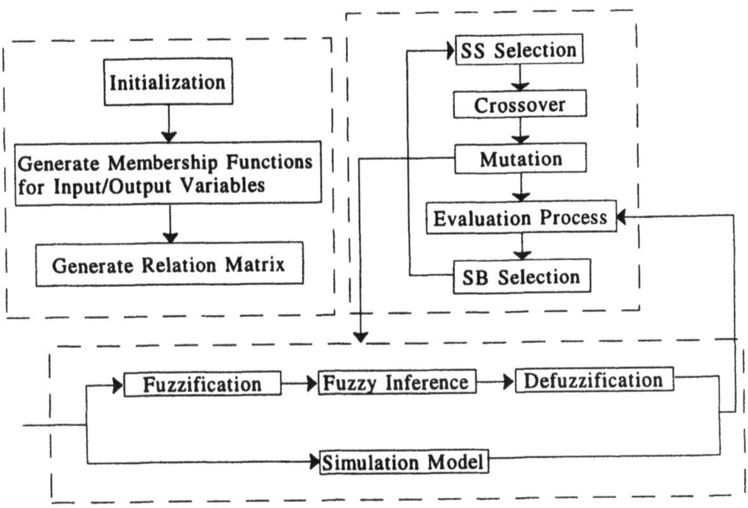

Fig. 12. Example of combining GA with fuzzy modelling (Zhou et al. 1997).

4 Conclusion

A common feature in Soft Computing methods is that they, in one way or another, mimic the human behaviour. Fuzzy logic has its power in describing vague and uncertain facts in a way that humans do in everyday life. Neural networks try to follow the structure of a human brain and their strength is in their ability of learning. Genetic algorithms have adopted their whole taxonomy from genetics and their strength is in using variety in optimisation.

All methods are based on sound mathematical premises. Their applications are increasing in numbers also in process industries and especially fuzzy logic control and neural networks applications in modelling have gained favour. This is also facilitated by the fast development of information technology that provides the applicationists with computer power and good possibilities for easy-to-use operator interfaces.

The following chapters describe some Finnish applications of Soft Computing. The 1990's saw a great boom of Soft Computing applications in Finland that was very much supported by the major research program "Adaptive and Intelligent Systems Applications" that was financed by National Technology Agency, Tekes. The program was conducted during 1994–1999 and 12 universities and 85 companies participated in the research projects and 110 companies in industrial or product development projects (Final Report 2000). The program produced a big number of real applications and prototypes and also some examples are shown here in this book.

References

Babuska, R., Verbruggen, H.B. (1995): A new identification method for linguistic fuzzy models. In: Proceedings FUZZ-IEEE/IFES'95, Yokohama, Japan, pp. 905–912

Valente deOliveira, J. (1993): Neuron inspired rules for fuzzy relational structures. Fuzzy Sets and Systems 57(1), 41–55

Driankov D., Hellendoorn H., and Reinfrank M. (1993): An introduction to fuzzy control. Springer, Berlin Heidelberg New York

Dubois, D., Prade, H. (1993): Fuzzy sets: A survey of engineering applications. Computers & Chemical Engineering 17, pp. S373–380

Final Report (2000): Adaptive and intelligent systems applications. Technology Program Report 18/2000. Tekes, Helsinki, Finland.

Glorennec, P. Y. (1994): Learning algorithms for neuro-fuzzy networks. In: Fuzzy Control Systems. CRC Press, Boca-Raton, Fl., pp. 3–18

Goldberg, D.E. (1989): Genetic algorithms in search, optimization, and machine learning. Addison-Wesley Publishing, Reading

Ishigami, H.; Fukuda, T.; Shibata, T.; Arai, F. (1995): Structure optimization of fuzzy neural network by genetic algorithm. Fuzzy Sets and Systems 71, 257-264

Huang, S.H., Zhang, H.-C. (1995): Neural-expert hybrid approach for intelligent manufacturing: A survey. Computers in Industry 26, 107-126

Jang, J.-S. R. (1992): Self-learning fuzzy controller based on temporal back-propagation. IEEE Transactions on Neural Networks 3(5), 714–723

Jang, J.-S.R. (1993): ANFIS: Adaptive-network-based fuzzy inference systems. IEEE Transactions on Systems, Man, and Cybernetics 23(3), 665–685

Juuso, E.K. (1996): Computational intelligence in distributed interactive synthetic environments. In: Bruzzone, A.G., Kerchoffs, E.J.H. (eds.): Simulation in process industry. 8th European Simulation Symposium. SCS, Ghent, Belgium, pp.157–162

Hunt, K.J., Sbarbaro, D., Zbikowski, R., Gawthrop, P.J. (1992): Neural networks for control systems – A survey. Automatica 28(6), 1083–1112

Kohonen, T. (1997): Self-organizing maps. 2nd Edition. Springer, Berlin Heidelberg New York

Lacher, R.C., Hruska, S.I., Kuncicky, D.C. (1992): Backpropagation learning in expert networks. IEEE Trans. Neural Networks 3(1), 62–72

Linkens, D.A.; Nyongesa, H.O. (1995): Genetic algorithms for fuzzy control, Part 1 offline system development and application. IEE Proc.-Control Theory Appl. 142(3), 161–176

Linkens, D.A.; Nyongesa, H.O. (1995): Genetic algorithms for fuzzy control, Part 2 Online system development and application. IEE Proc.-Control Theory Appl. 142(3), 177–176

Mamdani, E.H. (1974): Application of fuzzy algorithms for control of simple dynamic plant. Proc. IEE 121(12), 1585–1588

Michalewicz, Z. (1996): Genetic algorithms + data structures = evolution programs. 3rd Edition. Springer, Berlin Heidelberg New York

Ross, T. J. (1995): Fuzzy logic with engineering applications. McGraw-Hill, New York

Rummelhart, D.E., Hinton, G.E., Williams, R.J. (1986): Learning internal representations by error propagation. In: Parallel Data Processing. M.I.T. Press, Cambridge, MA, pp. 318–362

Takagi, T., Sugeno, M. (1985): Fuzzy identification of systems and its application to modeling and control. IEEE Transactions on Systems, Man and Cybernetics 15(1), 116–132

Verbruggen, H.B., Bruijn, P.M. (1997): Fuzzy control and conventional control: What is (and can be) the real contribution of fuzzy systems. Fuzzy Sets and Systems 90, 151–160

Wang, L-X. (1994): Adaptive fuzzy systems and control, design and stability analysis. Prentice Hall, New Jersey

Wells, G. (1995): An introduction to neural networks. In: Boullart, L., Krijgsman, A., Vingerhoeds, R.A. (eds.): Applications of Artificial Intelligence in Process Control. Pergamon Press, New York

Waterman, D.A. (1986): A guide to expert systems. Addison-Wesley Publishing Company, Reading

Zhou, J., Juuso, E., Leiviskä, K. (1997): Intelligent methods in peat production. In: Zimmermannn, H.-J. (ed.): EUFIT'97 – Fifth European Congress on Intelligent Techniques and Soft Computing Proceedings. Verlag Mainz, Aachen, Volume 3, pp. 2103–2107

Zadeh, L. A. (1965): Fuzzy sets. Information and Control 8(3), 338–353

Zadeh, L.A. (1975): The concept of a linguistic variable and its application to approximate reasoning. Part 1. Information Sciences 8, 199–249

Zadeh, L.A. (1994): Fuzzy logic and Soft Computing: issues, contentions and perspectives. In: Proceedings of IIZUKA '94. Third Int. Conf. On Fuzzy Logic, Neural Nets and Soft Computing, pp. 1–2

Soft Computing Applications in Mineral and Metal Industries

Kauko Leiviskä
University of Oulu, Control Engineering Laboratory, P.O.Box 4300, FIN-90014 Oulun yliopisto, Oulu, Finland

1 Introduction

Soft Computing has many industrial application areas, also in mineral and metal processing. Intelligent methods are used in Software Sensors to make the existing measurements more efficient or to replace the non-existing measurements with software systems that form the measurement signals e.g. from other, existing measurements. Both fuzzy logic and neural networks have been used. Another area is the monitoring of the measurement systems and assuring the good condition of the system.

In process control both open loop and closed-loop applications are in use. Advisory systems are based on process measurements and they are giving advises to process operators. In this case, only the operator closes the control loop. Basically, two variants of closed-loop control exist. First, Soft Computing Systems can replace existing controllers. Secondly, they are used as supervisory or adaptive controllers for changing set points or control strategies and adapting to varying operational conditions.

Diagnostics and fault detection systems are classical applications. Compared with algorithmic-based fault detection the biggest advantage is that they give possibilities to follow human's way of fault diagnosing and handle different information and knowledge in a more efficient way. Applications vary from troubleshooting of hydraulic and electronic systems to vibration analysis. The essential feature in the future is combining of Soft Computing methods with mathematical models and other conventional tools and finally their integration with total condition monitoring systems.

In production scheduling applications, systems are targeting to the higher throughput, the keeping of delivery times and good end-product quality. The functions vary from the allocation of customer orders to the selection of the routing in the mill according to product requirements. Two main possibilities are available (Leiviskä 1992): integrating Soft Computing with conventional algorithmic scheduling or replacing existing systems with new ones. Combined systems that utilise fuzzy logic, neural networks and genetic algorithms are used, because one approach alone can seldom lead to even feasible solution.

There are a lot of expert system applications in mineral and metal industries and in many areas they are already "conventional" technology. Neural network applications have increased recently. Schlang et al. (2000) report several hundreds of neural network applications in steel mills, only. Control systems produce lot of data and there are standard neural network procedures for handling it. Fuzzy logic applications are still on their way to more extensive applications. Also applications of genetic algorithms still wait their time.

2 Mining and Mineral Processing

Meech (1995) gives an extensive survey on the state-of-the art and future of intelligent methods in mining industry. He also gives an exhaustive list of existing references (for instance, McDermott et al. 1992; Eggert et al. 1994; Benford and Meech 1992; Lacouture and al. 1991; Edwards and Mular 1992; Poirier et al. 1993; Meech and Harris 1992; Rakocevic and Meech 1995 and Kumar et al. 1993). In mining industry, Expert Systems are used but they are poor at mathematical problem solving; they cannot learn and are usually not adaptable; generalisation and summarisation skills are essentially non-existent. Speed issues will also arise in real-time applications. The methodologies that permit adaptation are fuzzy logic and neural networks. For instance, (Ruokonen 1996; Laine et al. 1995) are describing the Expert System for ore-type classification that utilises neural networks (SOM). Fuzzy control of ventilation processes in coal mines has been reported by (Chojcan et al. 1995).

Another area in mining industry that lends itself to Soft Computing is the pit design and scheduling problems. They have been dealt with in two articles by (Thomas 1996a,b). Actually there are two problems connected: Ultimate Pit Determination and Extraction Sequencing (Scheduling). They are not independent, but should be solved together. Several optimisation methods have been used in solving the separate problems, as surveyed by Thomas. Of these, only Tolwinski and Underwood's approach (Tolwinski and Underwood 1992), genetic algorithms (Denby and Schofield, 1994,1995) and simulated annealing are capable of dealing the overall optimisation problem. Also here the combination of different approaches seems necessary.

SAG and FAG milling has been a conventional area for expert system applications because of the complicated relationships between required power in grinding and the mill load (Paris and Cipriano 1991; Perry and Hall 1994; Bourassa and Lanthier 1999). Early applications of neural network exist also for ball mills (Thibault et al. 1991; Flament et al. 1993).

Expert Control in flotation is already widely used (Cipriano et al. 1991; Lacouture et al. 1991). A newer application combining image processing and expert systems is presented later in this book. Cipriano et al. (1997) and van

Deventer et al. (1997) also report this kind of applications. Newer publications refer to the application of:

- Fuzzy logic control (Carvalho and Durao 1999; Osorio et al. 1999; Bergh et al. 1998; Carvalho and Durao 2000)
- Neural networks (Cortez and Durao 1995; Durao and Cortez 1995; Bartolacci and Bouajila 2000)
- Genetic algorithms (Karr 1995)

in modelling and control of the flotation plant.

The FLC has been developed for a rotary dryer and its control performance to compensate for the variations in the output moisture of solids has been examined by experiments with the pilot plant dryer in University of Oulu (Koskinen and Yliniemi 1996). Step changes in different input variables as in the input moisture of solids and in mass flow of solids have been made. FLC has been tuned off-line based on a trial and error procedure by changing the rules and the mapping of membership functions. Experiments show that FLC combined with the PI-controller is able to maintain the output moisture of solids as desired. Also fuzzy modelling with linguistic equations has been used to help the tuning of FLC (Koskinen et al. 1998). Also a hybrid neural/PI-controller has been tested and found able to compensate for disturbances in the input moisture. The neural network controller itself is a s. c. direct inverse controller that utilises the inverse process model. The network has been trained by back propagation algorithm in Matlab's Neural Network Toolbox. An adaptive fuzzy controller for this process will be introduced later on in this book.

The fluidised bed granulation process has been modelled using artificial neural networks (Rämet et al. 1996). The independent input variables were the inlet air temperature, the atomising air pressure and the amount of the binder solution. The responses used were mean granule size and granule friability. Several fully connected multi-layer neural networks were tested. The training algorithm was a modified back propagation algorithm designated as the Weigend weight eliminator.

The behaviour of the powder bed height inside the fluidised bed process is difficult to model and its automatic control using conventional methods is questionable. The behaviour of materials is uncertain and the granulation mechanisms are complex and unclear. The process is also highly non-linear and there are severe difficulties in measuring the bed height. The bed height control is, however, significant for the efficient operation of the whole process. This all makes it a good candidate for fuzzy logic control. A prototype fuzzy logic controller was developed and tested in a laboratory scale apparatus (Rämet et al. 1996).

Neural networks have also been applied in reconciling inconsistent process data (Aldrich and van Deventer 1994, 1995) and as a soft sensor for particle size estimation (Du et al. 1997).

3
Applications in Steel Mills

Intelligent applications are already found in several process areas in steel industry: coking plant, sintering, blast furnace, electric arc furnace, continuous casting and rolling mills. Applications exist also in millwide operations, in scheduling and quality planning. The following presentation shows some examples without trying to be an exhaustive one.

3.1
Coking Plant

Palmu (1998) has described the application of fuzzy logic control for coking plant control at Rautaruukki Oy's Raahe mill. Fuzzy logic is used in complementing the existing heating control model. It is beneficial especially in cases when rapid changes in energy requirements happen during the coking period. These may be due to coal blend change or change in the grain size of coal. Fuzzy logic controller corrects energy balance according to the average coke end temperature and coking time. Inclusion of fuzzy logic improved the reliability of the heating control system.

3.2
Sintering

Maki et al. (1991) give an example of fuzzy sintering control. It is based on the estimation of uniformity of burning in the pallet width direction using five rows of temperature sensors, four sensors in each row. The distribution of the burning rate in the width direction is controlled by changing the material charged density expressed as the cut volume. This cut volume is measured by five ultrasonic level sensors and controlled by five sub-gate actuators. The master loop measures the uniformity of burning and calculates the set points for the cut volumes. It uses a fuzzy controller with 28 rules. The cut volumes are controlled by a conventional PI-controller (slave). Fuzzy control improved burn rate uniformity. Ebner et al. (1997) give another application.

Sintering control is considered also in one chapter of this book.

3.3
Blast Furnace

Expert systems are extensively used at blast furnaces. One example is given in Alaraasakka et al. (1998). The operation of the expert systems is in principle based on the way experienced operators behave during process disturbances. It can be divided into three stages:

- Situation analysis
- Recognising the abnormal phenomena
- Recommending the required actions.

Classification of horizontal and vertical profiles in blast furnaces has been quite popular application, which utilises different neural network approaches (Otsuka et al. 1992; Takada et al. 1994; Iida et al. 1995; Angstenberger 1996, 1995; Rausch 1997; Alaraasakka et al. 1998; Lassus and Saxén 1998). Temperatures, concentrations and material shapes have been considered. The final target, in most cases, is the improvement of the heat level control, and therefore operators or an expert system further process the classifier results.

In one case (Alaraasakka et al. 1998) the neural network application for the above-burden probe searches the position and classifies the shape of 4.4 metres wide "top part" of the temperature distribution. Furthermore it classifies the shape of 0.93 meters wide temperature distribution from the both sides of total distribution. In the case of Dango probe, the neural network application classifies the shape of temperature, CO_2-, $CO+CO_2$- and CO/CO_2-distributions. It was found out that self-organising map suits well to the classification of the blast furnace gas distribution and blast furnace operators now use the applications.

One example of these applications is also included in this book.

3.4 Electric Arc Furnace

Electric arc furnace control has utilised both neural networks and fuzzy logic (NAC 1997; Siemens 1997; Bogdanoff et al. 1998). Neural networks are used mostly in modelling the non-linear three-phase systems. Systems using neural networks require computational power, because they are trained almost continuously (400 000 samples per second analogue input system). Fuzzy logic applications work mostly on process management level giving optimum voltage set points, material additions and correct end-point of smelting.

The first version of FLC reported in (Bogdanoff et al. 1998) was kept as simple as possible in order to decrease the number of fuzzy rules. The fuzzy logic controller optimises current and voltage set points and gives instructions for right time material additions. It also gives operational advises for operators and staff people. For the first basket there are four fuzzy controllers, one for all three electrodes and one for combining outputs of these three fuzzy controllers. Rules have been made in a situation where all electrodes are adjusted at the same time. Rule bases of fuzzy logic controllers of each electrode contain 34 rules in total. For the second basket there are also four fuzzy controllers, one for all three electrodes and one for combining outputs of these three fuzzy controllers. Rule bases contain 12 rules in total. The rule bases are not complete. The rule base for combining outputs contains 27 rules in total. The most important input variables of the electrode control are temperatures of water-cooled panels and their derivatives. Input variables include also the temperature from the combustion gas panel and rates of energy usage. Output variables are instructions for material additions and changing the voltage set points.

3.5 Continuous Casting

Applications in the continuos casting are few. Back propagation networks have been used in breakout forecasting (Hatanaka et al. 1993). Both neural networks and fuzzy logic have been used in mold level control with good results (Furukawa et al. 1992; Frank 1997; Siemens 1997).

3.6 Rolling Mill

Rolling mill applications are numerous. Complementing control strategies based on predictions and models mostly use neural networks. Siemens (1997) has used back propagation neural networks in pre-calculations using results from the previous processing together with mathematical models since 1994. Neural networks provide considerably better predicting capability, compared with mathematical models. Schlang et al. (2000) give a newer review. Applications are available for both cold and hot rolling mills starting from pre-calculation of rolling force to actual control of flatness and profile. Considerable savings are also reported (Portmann et al. 1995; Poppe et al. 1995).

Neural-fuzzy approach has found use in the shape (flatness) control of Senzimir mill (Funabashi and Maenda 1995). Neural networks classify plate shapes and fuzzy reasoning selects a suitable control strategy. Typically, the choice is between roll bending and shifting the position of first intermediate rolls.

Mefos (Leven 1995) has studied back propagation networks combined with mathematical models and simulation. The aim has been to provide mills with faster predicting tools for on-line applications. Applications cover both hot and cold rolling mills and rolling force and flatness predictions. Lu (1998) has also used neural networks in rolling load prediction in hot rolling. The accuracy of the roll force model was improved when applying neural networks in a Korean mill (Lee and Lee 2000). Vermeulen et al. (1997) used neural networks in predicting the temperature after the last finishing stands.

Steel industry uses camera-based systems for monitoring of surface faults. Neural networks are used in fault classification and rule based (expert) systems in final quality control. Two network structures have been reported in the literature: Adaptive Resonance Map (ARM) and back propagation (Dominigues et al. 1995; Dupond et al. 1993).

Fuzzy logic has been used in tuning of an expert knowledge-based fault classifier used for classifying defects on steel strips (Leiviskä 1996). In the system, the models of defect classes are tuned by using feature information obtained from real defect cases by an image processing system. The system makes the tuning stage of the classifier installation considerably faster. Diagnosis applications are coming more and more important in the future (Schlang et al. 2000).

4 Other Processes

A fuzzy logic controller has been in operation in the zinc roaster furnaces since 1994 with good results (Rauma et al. 2000). Furnaces are operated closer to their theoretical maximum temperature. Temperature variations are much smaller than with manual operation. Bergh et al. (2000) report on a soft sensor and fuzzy logic control of a slag cleaning furnace.

A rule-based expert system has been applied in fault diagnosis and control of the leaching process (Wu et al. 2000). The system is used in defining the optimum pH and target flow rates in the zinc hydrometallurgy plant.

The use of self-organising maps at a copper electrorefining plant has been considered in (Rantala et al. 2000). Combining SOM with principal component analysis was found to give the best classifying results when the effects of anode impurities to the physical quality of cathode copper were analysed. Self-organising maps have also been used in fault diagnosis of a copper smelter (Jämsä-Jounela et al. 2000).

Machine vision and multi-layer perceptron (MLP) neural networks were used in the control of the vacuum gas degassing process in the steel mill (Kämäräinen et al. 2000). The system classifies the degassing quality and estimates the hydrogen content of the liquid steel.

5 Conclusion

Expert systems are dominating the field. The reason for their success is the amount of earlier references and their integration with neural networks and fuzzy logic control. Neural network applications are in classification and modelling, but also in direct control. Data-intensive applications seem to be the most advantageous area for neural networks because of their processing capacity. Fuzzy logic is coming into the market, mostly in control applications and in combination with expert systems and neural networks. Genetic algorithms are not visible, yet. Their applications will certainly come in the future and in the area of optimisation.

The future is for hybrid applications. Real world problems require the combination of several methods (expert systems, neural networks, fuzzy logic, genetic algorithms and conventional modelling), because the systems are dealing with several classes of data and knowledge. The success comes from the wisdom of combining different methods and process know-how.

References

Alaraasakka, M., Leiviskä, K., Seppänen, M. (1998): Neural network based classification of blast furnace gas distribution. In: Proceedings TOOLMET '98 Symposium – Tool environments and development methods for intelligent systems. Control Engineering Laboratory, University of Oulu, Oulu, pp. 40–44

Aldrich, C., van Deventer, J.S.C. (1994): The use of connectionist systems to reconcile inconsistent process data. The Chemical Engineering Journal 54, 125–135

Aldrich, C., ven Deventer, J.S.C. (1995): Comparison of different artificial neural nets for detection and location of gross errors in process. Ind. Eng. Chem. Res. 34, 216–224

Angstenberger, J. (1996): Blast furnace analysis with neural networks. In: Bulsari, A.B. (ed.): International Conference on Engineering Applications of Neural Networks Proceedings (EANN '96), London, UK, 37–40

Bartolacci, G., Bouajila, A. (2000): Application of multivariable tools to mineral processing data analysis and modelling: Flotation case. In: Jämsä-Jounela, S.-L., Vapaavuori, E. (eds.): Future Trends in Automation in Mineral and Metal Processing, IFAC Workshop Preprints. Finnish Society of Automation, Helsinki, pp. 188–193

Benford, P., Meech, J.A. (1992): Advising flotation operators using a real-time expert system. Minerals Engineering 5(10–12), 1325–1331

Bergh, L.G., Achurra, G.H., Chacana, B. (2000): Coal rate injection control in a slag cleaning furnace. In: Jämsä-Jounela, S.-L., Vapaavuori, E. (eds.): Future Trends in Automation in Mineral and Metal Processing, IFAC Workshop Preprints. Finnish Society of Automation, Helsinki, pp. 183–187

Bergh, L.G., Yinatos, J.B., Leiva, C.A. (1998): Fuzzy supervisory control of flotation columns. Minerals Engineering 11(8), 739–748

Bogdanoff, A., Yliniemi, L., Leiviskä, K. (1998): Prototype fuzzy logic controller for electric arc furnace. In: Proceedings of TOOLMET '98 Symposium – Tool environments and development methods for intelligent systems. Control Engineering Laboratory, University of Oulu, Oulu, pp. 49–53

Bourassa, M., Lanthier, R. (1995): Grinding circuit optimization at Mine Doyon. In: 27th Annual Meeting of the Canadian Mineral Processors. CIM Publications, Montreal, Canada, pp. 294–310

Carvalho, M.T., Durao, F. (1999): Performance of a flotation column fuzzy controller. In: Mastorakis, N. (ed.): Computers and Computational Engineering in Control. World Scientific and Engineering Society Press, Athens, Greece, pp. 220–225

Carvalho, M.T., Durao, F. (2000): Supervisor fuzzy controller of a flotation column. In: Jämsä-Jounela, S.-L., Vapaavuori, E. (eds.): Future Trends in Automation in Mineral and Metal Processing, IFAC Workshop Preprints. Finnish Society of Automation, Helsinki, pp. 200–205

Chojcan, J., Czogala, E., Tyma, G. (1995): On fuzzy control of the ventilation process in the Longwalls with methane emissions in coal mines. In: EUFIT'95 – Third European Congress on Intelligent Techniques and Soft Computing Proceedings. Verlag Mainz, Aachen, Germany, pp. 1036–1040

Cipriano A., Guarini, M., Soto, A., Briceno, H., Mery, D. (1997): Expert supervision of flotation cells using digital image processing. In: Hoberg, H., von Blottnitz, H. (eds.):

Proceedings of the 20th International Mineral Processing Conference. GDMB Publications, Aachen, Germany, pp. 281–292

Cipriano, A., Medina, A., Martinez, C., Gonzalez, G. (1991): Supervisor and control strategy for a flotation plant. In: Proceedings IFAC Workshop on Expert Systems in Mining and Metals Processing. Finnish Society of Automation, Helsinki, Finland, pp. 281–285

Cortez, L., Durao, F. (1995): An expert neural network to control a mineral flotation process. In: 25th International Symposium on the Application of Computers and Operations Research in the Mineral Industry (APCOM) Proceedings. AusIMM Publications, Carlton, Victoria, pp. 431–436

Denby, B., Schofield, D. (1994): Open-Pit Design and Scheduling by Use of Genetic Algorithms. Trans (Section A: Mining Industry), IMM 103, A21–A26.

Denby, B., Schofield, D. (1995): The Use of Genetic Algorithms in Underground Mine Scheduling. In: 25th APCOM Proceedings. AusIMM Publications, Carlton, Victoria, pp 389–394

Van Deventer, J.S.J., Bezuidenhout, M., Moolman, D.W. (1997): On-line visualisation of flotation performance using neural computer vision of the froth texture. In: Hoberg, H., von Blottnitz, H. (eds.): 20th International Mineral Processing Congress. GDMB Publications, Aachen, Germany, pp. 315–326

Dominigues, S., Campoy, P., Aracil, R. (1994): A neural network based quality control system for steel strip manufacturing. In: Artificial Intelligence in Real Time Control, Proceedings, Valencia, Spain, pp. 185–190

Du, Y.G., del Villar, R., Thibault J. (1997): Neural net-based softsensor for dynamic particle size estimation in grinding circuits. International Journal of Mineral Processing 52(2–3), 121–135

Dupond, F., Odet, C., Carton, M., Alexandre, P. (1993): Automatic recognition and analysis of surface defects on pickled steel. In: SPIE 1993 Proceedings, Orlando, FL, USA, pp. 78–89

Durao, F., Cortez, L.A. (1995): Neural network controller of a flotation process. In: 19th International Mineral Processing Conference (IMPC) Proceedings. SME Publications, Littleton, Co., pp. 245–250

Ebner, J., Gould, L., Raml, H., Stiasny, H., Wasner, N., Fuerschuss H. (1997): Fuzzy logic control improves sinter plant performance at VA Stahl Linz. Steel Times International, January 1997, pp. 35–38

Edwards, R., Mular, A.L. (1992): An Expert System Supervisor of a Flotation Circuit. CIM Bulletin 84(959), 69–76

Eggert, J., Folinsbee, J., Benford, P. (1994). SAG Mill Control at Dome Mines Using an Expert System, In: 26th Canadian Mineral Processors Proceedings. CIM Publications, Montreal, Canada, pp.16–22

Flament, F., Thibault, J., Hodouin, D. (1993): Neural network based control of mineral plants. Minerals Engineering 6(3), 235–249

Frank, P.M. (1997): Implementation of a fuzzy logic based level control concept for continuous casting at Hüttenwerke Krupp Mannesmann. In: ERUDIT TC A Workshop: "Applications in Steel Industry"', Aachen, Germany, 14 p

Funabashi, M., Maenda, A. (1995): A hybrid and neural hybrid expert systems: synergetic AI. IEEE Expert, August 1995, 32–40

Furukawa, K., Amo, K., Watanabe, S., Itahishi, M., Kosakai, I. (1992): Development of CC mould level control system by neural network. In: Current advances in Materials and Processes, Japan.

Hatanaka, K., Tanaka, T., Kominami, H. (1993): Breakout forecasting system based on multiple neural networks for continuous casting in steel production. Fujitsu Scientific and Technical Journal, 29, 265–270

Iida, O., Ushijima, Y., Sawada, T. (1995): Application of AI techniques to blast furnace operations. Iron and Steel Engineering 72(10), 24–28

ISIS. 1997. http://www.ento.vt.edu/Facilities/ OnCampus/ISISInfo.html

Jämsä-Jounela, S.-L., Vapaavuori, E., Salmi, T., Gränbärj, M., Vermasvuori, M. (2000): Fault diagnosis system for the Outokumpu flash smelting process. In: Jämsä-Jounela, S.-L., Vapaavuori, E. (eds.): Future Trends in Automation in Mineral and Metal Processing, IFAC Workshop Preprints. Finnish Society of Automation, Helsinki, pp. 444–448

Karr, C.L. (1996): Strategy for adaptive process control for a column flotation unit. In: Ramani, R.V. (ed.): 26th International Symposium on the Application of Computers and Operations Research in the Mineral Industry (APCOM) Proceedings. SME Publications, Littleton, Co., pp. 303–307

Koskinen, J., Yliniemi, L. (1996): Fuzzy control of a rotary dryer, In: Proceedings of TOOLMET'96 Symposium – Tool environments and development methods for intelligent systems. Control Engineering Laboratory, University of Oulu, Oulu, pp. 205–217

Koskinen, J., Yliniemi, L., Leiviskä, K. (1998): Fuzzy modelling of a pilot plant rotary dryer. In: International Conference on Control '98 Proceedings. University of Wales, Swansea, UK, 515–518

Kumar, S., Meech, J.A., Samarasekera, I.V., Brimacombe, J.K. (1993): An expert system to trouble-shoot quality problems in the continuous casting of steel billets. Iron & Steelmaker 20(9), 29–36.

Kämäräinen, J.K., Kälviäinen H., Terho, K., Saarelainen E. (2000): Visual quality control of the vacuum tank degassing. In: Jämsä-Jounela, S.-L., Vapaavuori, E. (eds.): Future Trends in Automation in Mineral and Metal Processing, IFAC Workshop Preprints. Finnish Society of Automation, Helsinki, pp. 379–384

Lacouture, B, Russell, C., Griffin, P., Leung, K. (1991): Copper Flotation Expert System at Mount Isa Mines Limited. In: Dobby, G.S., Argyropoulos S.A., Rao, S.R. (eds.) Proceedings Copper 91, pp. 429– 437

Laine, S., Lappainen, H. Jämsä-Jounela, S.-L. (1995): On-line determination of ore type using cluster analysis and neural networks. Minerals Engineering 8(6), 637–648

Lassus, L., Saxén, H. (1998): System for pattern recognition of blast furnace wall temperatures. In: Proceedings of TOOLMET '98 Symposium – Tool environments and development methods for intelligent systems. Control Engineering Laboratory,, University of Oulu, Oulu, pp. 45–48

Lee, D., Lee, Y. (2000): Application of neural network for improving accuracy of roll force model in hot-rolling mill. In: Jämsä-Jounela, S.-L., Vapaavuori, E. (eds.): Future Trends in Automation in Mineral and Metal Processing, IFAC Workshop Preprints. Finnish Society of Automation, Helsinki, pp. 240–244

Leiviskä, K. (1992): Application Viewpoints of Expert Systems in Mineral and Metal Processing. In: Proceedings of the IFAC Workshop Expert Systems in Mineral and Metal Processing. Pergamon Press, Oxford, pp. 191–195

Leiviskä, K. (1996): Applications of intelligent methods in the Control Engineering Laboratory, University of Oulu. In: Proceedings of TOOLMET '96 Symposium – Tool environments and development methods for intelligent systems. Control Engineering Laboratory, University of Oulu, Oulu, pp. 28–34

Leven, J., Jonsson, N.G., Wiklund, O. (1995): An artificial neural network for rolling applications. Steel Times 223(4), 137–138

Lu, C. (1998): Application of ANN in a combination with mathematical model in prediction of rolling load of the finishing stands in HSM. In: 7th International Conference on Steel Rolling Proceedings, Chiba, Japan, pp. 87–92

Maki, Y., Takashima, N., Obata, H., Iida, O., Nakahima, K., Sawada, T. (1991): Application of fuzzy theory to ironmaking process control. Kawasaki Steel Technical Report, No. 24, pp. 8–10

McDermott K., Clyle P., Hall, M., Harris, C.A. (1992): An Expert System for Control of No.4 Autogenous Mill Circuit at Wabush Mines. In: Canadian Mineral Processors Proceedings. CIM Publications, Montreal, Canada, Paper 24, pp.20-24

Meech J.A. (1995): AI Applications in the Mining Industry into the 21st Century. In: 25th International Symposium on the Application of Computers and Operations Research in the Mineral Industry (APCOM) Proceedings. AusIMM Publications, Carlton, Victoria. Also in: (http://www.mining.ubc. ca /faculty/meech/apcom.htm)

Meech J.A. and Harris, C.A. (1992): Expert Systems for Gold Processing Plants. In: Randol Gold Forum, Vancouver, B.C., pp. 45–61

Osorio, P., Perez-Correa, J.R., Cipriano, A. (1999): Assessment of expert fuzzy controllers for conventional flotation plants. Minerals Engineering 12(11), 1327–1338

Otsuka, Y., Matsuda, K., Hanaoka, K. (1992): Application of neural network systems to pattern recognition of blast furnace operation data. Kobleco Technology Review, No.15, 12–16

Paris, A., Cipriano, A. (1991): Supervisor Expert Control of a semi autogenous grinding circuit. In: IFAC Workshop on Expert Systems in Mineral and Metals Processing. Finnish Society of Automation, Helsinki, pp. 211–216

Palmu, P. (1998): Experience of fuzzy logic control system at Rautaruukki Raahe Steel. In: Proceedings of TOOLMET '98 Symposium – Tool environments and development methods for intelligent systems. Control Engineering Laboratory, University of Oulu, Oulu, pp. 32–39

Perry, R., Hall, M. (1994): Les Mines Selbaie semi-autogenous grinding circuit expert system. In: 26th Annual Meeting of the Canadian Mineral Processors Proceedings. CIM Publications, Montreal, Canada, pp. 294–310

Poirier, P., Raabe, H., Meech, J.A. (1993): Using froth identification in an advisory expert system for copper flotation operations. In: 23th Canadian Mineral Processors Conference Proceedings. CIM Publications, Montreal, Canada, pp.14–19

Poirier, P.J., Meech, J.A. (1993): Using fuzzy logic for on-line trend analysis. In: 2nd IEEE Conf. on Control Applications Proceedings, Vancouver, B.C., pp. 83–86

Poppe, T., Schlang, M., Obradovic, D. (1995): Neural network: reducing energy and material requirements. Siemens Review 4, 24–27

Portmann, N.F., Lindhoff, G., Sörgel, G., Gramckow, O. (1995): Application of neural networks in rolling mill automation. Iron and Steel Engineering 72(2), 33–36

Rakocevic, V.M., Meech, J.A. (1995): An artificial neural network to interpret froth images from a flotation process. CIM Bulletin 88, 16–22

Rantala, A., Virtanen, H., Saloheimo, K., Jämsä-Jounela, S.-L. (2000): Using principal component analysis and self-organising map to estimate the physical quality of cathode copper. In: Jämsä-Jounela, S.-L., Vapaavuori, E. (eds.): Future Trends in Automation in Mineral and Metal Processing, IFAC Workshop Preprints. Finnish Society of Automation, Helsinki, pp. 373–378

Rauma, T., Nyberg, J., Herranen, J. (2000): Fuzzy control of furnace temperature in a zinc roaster. In: Jämsä-Jounela, S.-L., Vapaavuori, E. (eds.): Future Trends in Automation in Mineral and Metal Processing, IFAC Workshop Preprints. Finnish Society of Automation, Helsinki, pp. 161–165

Rausch, H. (1997): Control of blast furnace performance using neural networks. In: ESIT '97 – First European Symposium on Applications of Intelligent Technologies Proceedings. Verlag Mainz, Aachen, Germany, pp. 37–44

Ruokonen, E. (1996): Application of the expert system based on neural network at the Kemi concentrator. In: Proceedings of TOOLMET'96 Symposium – Tool environments and development methods for intelligent systems. Control Engineering Laboratory. University of Oulu, Oulu, pp. 102–110

Rämet, P.; Lehtola, S.; Leiviskä, K.; Mannermaa, J-P. (1996): The fluidized bed granulation process modelling and control using neural networks and fuzzy control. In: Zimmermannn, H.-J. (ed.): EUFIT'96 – Fourth European Congress on Intelligent Techniques and Soft Computing – Proceedings. Verlag Mainz, Aachen, Germany, pp. 1924–1928

Schlang, M., Jansen, M., Lang, B., Poppe, T., Runkler, T., Weinzierl, K. (2000): Current and future developments in neural computation in steel processing. In: Jämsä-Jounela, S.-L., Vapaavuori, E. (eds.): Future Trends in Automation in Mineral and Metal Processing, IFAC Workshop Preprints. Finnish Society of Automation, Helsinki, pp. 27–37

Siemens: Newsletter Metals, Mining and More, Edition 3/97. Available on-line at: http://www.atd.siemens.de/metals_mining_more/index.html. 25.9.1998

Takada, H., Hamada, K., Kaya, K. (1994): A design environment for neural networks and its applications. Control Engineering Practice 2(2), 123–128

The intelligent arc furnace controller. Available on-line at http://www.neural.com /IAFController/IAFController.htm. 16.4.1998

Thibault, J., Flament, F., Hoduin, D. (1991): Modelling and control of mineral processing plants using neural networks. In: IFAC Workshop on Expert Systems in Mineral and Metals Processing. Finnish Society of Automation, Helsinki, pp. 281–288

Thomas, G.S. (1996a): Optimisation and scheduling of open pits via genetic algorithms and simulated annealing. In: Proceedings of the 1st International Symposium on Mine Simulation via the Internet in December 1996

Thomas, G.S. (1996b): Pit optimisation and mine production scheduling – The way ahead. In: Ramani, R.V. (ed.): 26th International Symposium on the Application of Computers and Operations Research in the Mineral Industry (APCOM) Proceedings. SME Publications, Littleton, Co., pp 221–228

Thomson, S.J., Peart, R.M. (1986): An expert system for soil moisture-based scheduling of center pivot irrigation. ASAE. Paper # 86, 4518

Tolwinski B., Underwood R. (1992): An algorithm to estimate the optimal evolution of an open pit mine, In: The 23rd International Symposium on the Application of Computers and Operations Research in the Mineral Industry (APCOM) Proceedings. SME Publications, Littleton, Co., pp 399–409

Vermeulen, W., Bodin, A., Zwaag, S. (1997): Prediction of measured temperature after the last finishing stands using artificial neural networks. Steel Research 68(1), 20–26

Wu, M., She, J.H., Nakano, M., Gui, W. (2000): Expert control and fault diagnosis of the leaching process in zinc hydrometallurgy plant. In: Jämsä-Jounela, S.-L., Vapaavuori, E. (eds.): Future Trends in Automation in Mineral and Metal Processing, IFAC Workshop Preprints. Finnish Society of Automation, Helsinki, pp. 351–356

Soft Computing Applications in Pulp and Paper Industry

Kauko Leiviskä
Control Engineering Laboratory, University of Oulu, P.O.Box 4300, FIN-90014 Oulun yliopisto, Oulu, Finland

1 Introduction

In Scandinavia, paper industry has been in the pioneering role in the development of process automation that started already in the 1950s with centralised control rooms and standardised signal systems. Computerised process automation dates from the early 1960s with paper machine and digester control systems, when business computers of that time were the first computer control systems (Leiviskä 1999a).

The development continued during the 1970s with dedicated process computers and their applications, so called packaged systems, to the control of different processes of pulp and paper mills. This development was made possible because of new instrumentation and specialised control system vendors.

The first digital automation systems came to the market at the end of the 1970s meaning the change in the application environment and offering new possibilities to the control development. Modern process automation takes the advantage of the distribution of control functions and centralising process monitoring.

During the 1990s systems integration continued on mill and corporate levels, together with the integration with office automation, personal data processing and general networks, mostly meaning Internet. This kind of application environment has also made the building of intelligent knowledge-based systems very natural, computer tools are available and also system vendors have adopted intelligent systems in their palette of tools. This is recognised, e.g. by Fadum (1996), Finchem (1998), and (Leiviskä 1999b).

2 Fibre Line Applications

2.1 Digester House Applications

Digester house is the primary production unit in the pulp mill and the successful quality control is the key to good paper quality. This has also set the direction to applications of soft computing in the digester; modelling and control of the quality variable, Kappa number, is of the primary importance.

With a good Kappa number estimate, digester control can be greatly improved. By measuring the liquor concentrations from the quench circulation of the impregnation vessel, the transfer circulation after the impregnation vessel, trim circulation and extracting black liquor, Kappa number can be estimated in different stages of the continuous cooking process. The first approach was to use neural networks to solve this estimation problem (Haataja et al. 1997). Later Linguistic Equations Approach was used in the same problem with good results (Murtovaara et al. 1998). Using wavelets in data pre-processing improves modelling, because data is more stable. This way data corresponds better the real measurements than by using the median filter (Murtovaara et al. 1998). A comparison of different modelling approaches in this case is presented later in this book.

Stronger pulp, fewer rejects and lower consumption of raw materials and energy has resulted from the fuzzy logic control of the digester house (Furumoto 1995). In the same application, neural networks are used in defining the precise end point of the cook, which further decreases the variations in Kappa number.

2.2 Washing Plant

Washing plant has dual objectives in the pulp mill: to produce pure pulp without used chemicals and collect chemicals as completely as possible to be recovered in the recovery processes. These objectives are partly opposite and optimisation is needed.

Rudd (1991) reports on the application of neural networks to the brown stock washer. Neural networks have use as soft sensors to measure mat consistency, mat density, and soda loss, and to control the dilution factor. The target is to stabilise the black liquor solids. Results show a 25% reduction in the standard deviation of the black liquor solids during an eight-day trial. The controller also maintained larger disturbances in an automatic mode for the input variables. Wilks (1997) reports on application of an expert system to optimise the operation of brown stock washing at an Australian mill.

2.3 Bleach Plant

In the bleach plant control, quality comes to the picture once again. Bleaching must produce pulp with good and constant quality still minimising the bleaching costs,

Fuzzy logic control of D_{100}-stage bleaching is reported in (Lampela et al. 1996). Several reasons make fuzzy logic suitable for this purpose. The system is non-linear and controller tuning depends on the operation point. Raw material is heterogeneous. There are severe interconnections between process variables and several variables effect on the optimum chemical dosage. Continuous measurements, laboratory analyses and human knowledge are easy to incorporate in the strategy.

The application is combined feedforward/feedback as in many pulp mill cases. The actual control output is the set point to the dosage controller. The standard deviation of EOP Kappa decreased by 60% and the standard deviation of the final brightness by 40% during a three-week test period after installing the fuzzy logic based system in conventional five-stage bleaching in Finland.

Diaz et al. (1992) used the chlorination stage as an example of using qualitative modelling techniques in process control and diagnostics. Neural networks have also been applied to bleach plant modelling and control (Xia et al., 1993; Zhu et al. 1997). The system integrates data preparation, the selection of the network topology and training to simulation. Koepenick (1996) has commented on considerable savings gained by fuzzy control at Caima mill in Portugal. Kosunen et al. (1996) report on the use of fuzzy logic and neural networks in ECF/TCF bleach plant. Another article tells about using the same technology in controlling the oxygen delignification stage in the pulp mill in Finland (Pulkkinen et al. 1997). Combining on-line Kappa analyser and fuzzy logic for the control of oxygen delignification has been considered also in (Ollila 1997).

3 Applications in Chemical Recovery

3.1 Evaporation Plant and Recovery Boiler

Evaporation and recovery boiler are the main processes in chemical recovery. In evaporation, the solids content of used chemicals in water solution is increased to the level economic in burning. The solids are burnt in the recovery boiler that, on one hand, recovers the chemicals and, on the other, guarantees the steam production for pulp mill processes.

The performance of a recovery boiler is strongly affected by the geometry of the char bed, and therefore the operation of the boiler can be improved, if this geometry is known. By using infrared fire-room cameras, the bed can not only be displayed to the operator but it is also possible to calculate the geometry parameters of the char bed, when the edges of the bed have been detected. A remarkable improvement potential exists in utilising digital image processing on fire-room camera information and in this way expanding and improving burning supervision and control. An example of the expert system approach is shown in Hosti et al. (1994).

A fuzzy edge detection algorithm was developed in order to improve the performance of the image processing system of the recovery boiler (Murtovaara et al. 1996). The image processing system tries to find out the contour of the bed. A new algorithm is based on the membership functions of the contour obtained from history data. This algorithm filters out fast changes of the contour and takes into account the neighbouring points by examining the new contour, and if the distance between the forecasted pixel and the pixel obtained by image processing goes over tuneable limits, the algorithm removes these pixels. This further improves its efficiency and increases the accuracy.

Ozaki et al. (1994) have described a system where the bed shape from the image analysis system is first classified by a three-layer back propagation neural network into three classes. The fuzzy controller changes the air flows into the recovery boiler by using the classifier information.

Good results of using the fuzzy logic control in the symmetry control are available in Mikkonen et al. (1994) and in Lampela et al. (1997). Also Nakatani (1995) report results from the fuzzy logic control of 1750 tpd recovery boiler.

Smith and Edwards (1991) propose the combination of simulation and a real time expert systems to the recovery boiler control. The recovery boiler model uses a steady-state modular simulator. A real time advisory control integrates process knowledge in the form of rules and simulated values with process data. Rules use statistical process control (SPC) methods to reveal when a variable is in abnormal condition. This makes easily understood detection of sensor problems and process upsets possible in real time. Horton et al. (1998) give an example of using neural networks in modelling the operation of the recovery boiler.

Brooks and Edwards (1992) report an application of the real time expert system for the evaporation plant control. It includes the simulation and calculation of heat transfer coefficients and the advisory system.

3.2
Causticisation and Lime Kiln

The first lime kiln control systems based on fuzzy logic was installed in a Swedish paper mill already at the end of 1979 (Ostergaard 1993). Other installations followed later using a system that aims to stabilise and optimise lime quality. It also minimises fuel consumption and optimises production. The control itself includes the burning zone and combustion controls. Good results were reported (Nilsson and Langkjaer 1996). Variations in residual carbonate have decreased by two-thirds. The supply of makeup lime has decreased from 10% to 5%. Fuzzy logic is also controlling the lime addition in the slaker resulting 1–2 % increase in the causticising efficiency (Yeager 1997). Nilsson and Dobel (1998) have also reported considerable savings with the fuzzy logic control of lime kiln and causticisation plant at a Swedish mill.

A fuzzy expert system application for the lime kiln control was reported in Leiviskä et al. (1988). Fuzzy logic control of the flue gas fan in the lime kiln was reported by (Haataja and Ruotsalainen 1994). FLC was originally implemented in the mill's automation system. It had seven inputs and one output. The rule base consisted of 22 rules that were formulated originally based on operator interviews and later tuned manually by trial and error. The results were encouraging; the annual savings in energy approached 2 million Finnish Marks. The modelling of the burning temperature is a part of a larger study, which aims to further improve the fuzzy logic control of the lime kiln (Juuso et al. 1997). The results are used in designing the burning temperature control for the kiln and in off-line testing of control alternatives. This system is described in more details later in this book.

The internal model control strategy for the lime kiln control has been suggested (Ribeiro et al. 1993). It uses multi-layer, feedforward back propagation network with 8 inputs, 2 outputs, and 2 hidden layers with 20 and 10 nodes.

3.3 Thermomechanical Pulping

In thermomechanical pulping there is a trade-off between pulp quality and the use of electrical energy. Good quality pulp must be produced with the minimum energy consumption. Also the development in quality analysers has made the application of advanced control methods possible.

The fuzzy control strategy was developed for a TMP-plant based on the freeness and mean fibre length controls with the specific energy consumption and using on-line freeness and fibre length analysers (Myllyneva et al. 1997). Two fuzzy controllers are used in a cascaded mode. The first controller observes the fibre length and smoothly begins to raise the freeness set point if the fibre length begins to go too short. The controller checks if the fibre length allows reduce the freeness set point. The second controller keeps freeness at the desired level by changing the set point of the total specific energy. Also this application is commented later.

4 Paper Mill Applications

Paper mill is dealing with the final product. Therefore, both requirements of high production efficiency and product quality are emphasised. The applications of Soft Computing methods vary from process control to diagnostics and from quality monitoring and control systems to production scheduling.

A prototype web break sensitivity indicator was developed to give the process operators a continuous indication of web break sensitivity in an easily understandable way (Ahola et al. 1999). It is also important to the operators to get support in finding out the reason of the break to be able to avoid it in the future. Being able to indicate the break risk and to avoid breaks has two major benefits: the machine efficiency is increased directly and fewer disturbances in the process improve process stability and efficiency. The prototype of web break sensitivity indicator has been implemented as a Case-Based Reasoning-type application with Linguistic Equations Approach and fuzzy logic. The indicator contains 25 separate models. It compares on-line data to these models in database and estimates how well the present measurements fit to the equations. The first on-line tests in practice have been done in a Finnish paper mill.

A back propagation algorithm has been used in diagnosing web breaks in 400 tpd Japanese mill (Miyahishi and Shimada 1997). A neural network system for fault classification in web inspection in described in (Oyama et al. 1997).

Also Fadum (1996) gives examples of using expert systems in diagnosing the reasons for paper breaks. Expert systems have also been used in personnel training and diagnostic tools for saw mils, wood yards, paper machines and wastewater treatment plant (Morin and Coderre 1996).

Dryer Expert System has been installed at Domtar Fine Papers in Canada to generate information through diagnosis of dryers and press section and to help operators to bring the operation back to normal conditions. The system uses 120 process inputs (Nault and Maltais 1997). Amyot et al. (1998) also report on the application of the on-line expert system for troubleshooting and optimisation of the paper machine dryer section in another Canadian mill.

An expert system utilising SPC has been used to control a coating line (Woollard et al. 1996). The system uses on-line data of key process parameters, tries to identify causes of out-of-control conditions and helps the operator to correct the situation.

Soft Computing methods are also used in process analysis. Powell (1996) gives some examples of using neural networks in some paper mill cases. Neural networks have also been used in prediction of paper curl (Edwards et al. 1999), CMT and porosity values for the entire period of reel production (Rudd 1995), smoothness (Hergert and Ahner 1999) and optical properties of paper; brightness and opacity (Musavi et al. 1999). The latter case uses radial basis functions together with multi-resolution analysis and fast-orthogonal search for training. Operation data from a deinking plant was analysed using neural networks (Stapley et al. 1995; Broeren et al. 1995). Results were used in optimising the plant and \$150 000 – \$289 000 annual savings were reported. Neural networks have been used in modelling and simulation of paper forming processes (Scharcanski and Dodson 1995). Standard feedforward networks and radial basis function networks are compared in modelling the influence of process variables in the chemical treatment of jack pine (Lanouette et al. 1997).

Nyborg and Crowell (1997) describe a simulation software and an expert system for analysing different paper quality variables based on furnish properties and machine conditions. Factor analysis and fuzzy modelling is described in Zhang et al. (1997) for modelling of quality of non-wovens and polymeric composites.

An expert diagnostic system for the predictive maintenance in the paper machine press section is described in (Sotelo et al. 1996). The system uses a dynamic simulation model of vibrations in the press section of the paper machine. Neural networks are also used in vibration analysis and predictive maintenance in paper machines (Pringle 1992).

Expert Systems are also widely used in production scheduling. Mori et al. (1997) describe an application in a Japanese paper mill. The mill has ten paper machines and it produces more 200 different production runs on a monthly basis. A new scheduling system decreased working power and simplified management operations.

A three-stage algorithm has been developed to help basis weight profile control (Sasaki 1996). It includes a competitive learning algorithm, wavelet transformations and a back propagation algorithm. Neural networks have also been used in identification and tuning of the CD controllers in the paper machine (Samad et al. 1998). Expert Systems have also been developed for monitoring the control loop performance.

5 Other Applications

Several applications have been reported in forestry (Brack and Marshall 1996; Chiu and Occena 1995), and land usage (Hsieh et al. 1997; Ells et al. 1997; Basri and Stentiford 1997), in energy management (Nelson 1997; Baines et al. 1997), and other areas (Mikkonen 1996; Jordan et al. 1997; Suojärvi 1997; Malmkvist et al. 1996; Back et al 1997).

6 Conclusions

There is a growing number of applications of Soft Computing in Pulp and Paper Industry varying from expert systems to fuzzy logic, neural networks and genetic algorithms. Based on the experiences from several industrial case studies, these methods seem to have benefits in several areas and uses. Development seems to go to the direction of data pre-processing that needs more attention in all application fields and hybrid methods combining the advantages of fuzzy logic, neural networks and genetic algorithms are winning more room in the future.

In control applications, measurement problems and methods for interpreting results of measurements seem to be of primary importance. This is clearly seen in the digester and recovery boiler cases handled before. Also conventional fuzzy logic control can give considerable savings, as in the bleach plant and TMP cases. Soft Computing leads to advantages in complicated data-intensive process analysis and diagnosis.

References

Ahola, T., Ruuska, J., Juuso, E., Leiviskä, K. (1999): Web Break Sensitivity Indicator. In: Proceedings of TOOLMET'99 Symposium – Tool Environments and Development Methods for Intelligent Systems. University of Oulu, Oulu, pp. 202–209

Amyot, R., Yeung, L., Phan, S., Henzell, R., Futcher, J., Kricorissian, K., McLellan, F. (1998): Papermaker's advisor. In: 84th Annual Meeting Preprints. Technical Section CPPA, Montreal, pp. B177–179

Back, B., Irjala, M., Sere, K., Vanharanta, H. (1997): Competitive financial benchmarking using self-organising maps. Pap. Puu 79(1), 42–49

Baines, G.H., Hayes, R.L., Stabell, J.L. (1997): Prediction boiler emission with neural networks. Tappi J. 80(5), 57–61, 87, 95, 103

Basri, H., Stentiford, E.I. (1997): Expert system for environmental monitoring of sanitary landfills. In: 1997 Environmental Conference Proceedings. Tappi Press, Atlanta, pp. 381–384

Brack, C.L., Marshall, P.L. (1996): Test of knowledge-based forest operation scheduling procedures. Can. J. For. Res. 26(7), 1193–1202

Broeren, L.A., Smith, B.A., Conte, T.O., Stapley, C. (1995): Deinking process optimisation using neural networks. In: Proceeding of Wastepaper VI Conference. Paper no. 14, 25 p

Brooks, T. R., Edwards, L. L. (1992): A real-time, model-based expert system for multiple effect evaporators. Tappi J. 75(11), 131–135

Chiu, C.C., Occena, L.G. (1995): Neural models for characterization and interpretation of hardwood internal defect configuration. In: Proceedings of the 1995 Artificial Neural Networks in Engineering. ASME Press, pp. 905–910

Diaz, A. C., Orchard, R. A., Amyot R., Braham, J.W. (1992): Qualitative modelling techniques for process control and diagnosis. *Tappi* J. 75(11), 149–155

Edwards, P.J., Murray, A.F., Papadopoulos, G., Gordon, M.F., Wallace, A.R., Barnard, J. (1999): Paper curl prediction. IEE Conf. Publ 1, (470), 335–340

Ells, A., Bulte, E., van Kooten, G. (1997): Uncertainty and forest land use allocation in British Columbia: Vague priorities and imprecise coefficients. For. Sci. 43(4), 509–520

Fadum, O. (1996): Introduction to artificial intelligence. In: Fadum, O. (ed.): Process, Quality and Information Control in the Worldwide Pulp and Paper Industry. Miller Freeman Inc., New York, pp. 16–17

Fadum, O. (1996): Expert systems in action. In: Fadum, O. (ed.): Process, Quality and Information Control in the Worldwide Pulp and Paper Industry. Miller Freeman Inc., New York, pp. 23–34

Finchem, K. (1998): Neural network technology moves toward mainstream control use. Pulp Pap. 72(4), 63–64, 67–68

Furumoto, H. (1995): Advanced control strategies for digester: analytical model-fuzzy logic-neural networks. In: 28° Congresso Anual de Celulose e Papel. ABTCP, Sao Paolo, pp. 241–251

Haataja, K.; Leiviskä, K.; Sutinen, R. (1997): Kappa number estimation with neural networks. In: Proceedings of XIV IMEKO World Congress: New measurements – Challenges and Visions. Finnish Society of Automation, Helsinki, Volume XA, pp. 1–5

Haataja, K., Ruotsalainen, J. (1994): Fuzzy Logic in Lime Kiln Control. In: Proceedings, MEPP'94 (Reports on Computer Science & Mathematics, Åbo Akademi, Ser. B, No 17)

Hergert, D., Ahner, R.J. (1999): Using neural networks to improve smoothness on a paper coating process. In: Dagli, C. (ed.): Intelligent Engineering Systems through Artificial Neural Networks vol. 9. ASME Press, 727–732

Horton, R., Denlinger, M., Lien, S., Schmidl, W., Grace, T. (1998): Artificial neural network modeling of a kraft recovery boiler. In: 1998 International Chemical Recovery Conference Proceedings. TAPPI Press, Atlanta, pp. 283–298

Hosti, T., Leiviskä, K., Sutinen, R. (1994): Burning Expert System for Recovery Boilers, In: 1994 Control Systems Preprints. SPCI, Stockholm, Sweden, pp. 313–319

Hsieh, H., Cheng, C., Chen, Y. (1997): Building expert system for forest-land classification. Taiwan J. For. Sci., 3, 255–268

Jofriet, P., Seppala, C., Harvey, M., Surgenor, B., Harris, T. (1996): Expert System for control-loop performance. Pulp Pap. Can. 97(6), 70–74

Jordan, R., Feeney, F., Evertsen, J.A. (1997). Classification of wood species by neural network analysis of ultrasonic signals. Ultrasonics 36(1–5), 219–222

Juuso, E..K., Ahola, T., Leiviskä, K. (1997). Fuzzy Modelling of a Rotary Lime Kiln Using Linguistic Equations and Neuro-fuzzy Methods. In: Foulloy, L. (ed.): 3rd IFAC Symposium on Intelligent Components and Instruments for Control Applications (SICICA'97) Preprints. IFAC, Annecy, France, pp. 579–584

Koepenick, M. (1996): Clear results from fuzzy logic. PIMA 78(3), 38–41

Kosunen, J., van Fleet, R., Forslund, O. (1996): How can new control methods improve pulp-mill performance. In: 5th International Conference on Available New Technologies. SPCI, Stockholm, pp. 285–304

Lampela, K., Mikkonen, J., Prantner, M., Leiviskä, K. (1997): Experiences of the fuzzy logics in pulp mill. In: 51st Appita Annual General Conference Proceedings. Appita, Carlton, Victoria, Australia, pp. 527–534

Lampela, K., Kuusisto, L., Leiviskä K. (1996): D_{100}-stage bleaching control with fuzzy logic. Tappi 79(4), 93–97

Lanouette, R., Thibault, J., Valede, J.L. (1997): Improvement of Modelling D-optimal design data with neural networks. In: 83rd Annual Meeting Preprints. Technical Section CPPA, Montreal, A1–7

Leiviskä, K. (1999a): Control Systems. In: Leiviskä, K. (ed.): Process Control. Papermaking Science and Technology, Book 14. Fapet Oy, Jyväskylä, Finland, pp. 13–17

Leiviskä, K. (1999b): Control Methods. In: Leiviskä, K. (ed.): Process Control. Papermaking Science and Technology, Book 14. Fapet Oy, Jyväskylä, Finland, pp. 19–47

Leiviskä, K., Huttunen, R., Uronen, P. (1988): Expert Systems in the lime kiln control. In: 1988 Conference on The Use of Computers in Chemical Engineering (CHEMDATA 88) Proceedings 2. EFCE, Gothenburg, p. 379

Malmkvist, K., Verikas, A., Malmkvist, L., Bergman, L. (1996): New method for color measurements in multicolor newspaper pictures and its use for printing control. In: International Printing and Graphic Arts Conference Proceedings. Tappi Press, Atlanta, pp. 181–186

Mikkonen J., Fuzzy logic dips into active sludge. Pulp Pap. Eur. 1(9), 26–27

Mikkonen, J., Koskinen, J., and Hanhinen, J. (1994): Fuzzy control in the recovery boiler. Paperi ja Puu 76(5):5

Miyahishi, T., Shimada, H. (1997): Neural networks in web-break diagnosis of a newsprint paper machine. In: 1997 Engineering & Papermakers Conference Proceedings. Tappi Press, Atlanta, pp. 605–612

Mori, Y., Nishimura, M., Chujima, T., Yokoyama, H., Abe, T., Yasuda, Y., (1997): Paper production scheduling system using an expert system. Tappi J. 80(11), 121, 129, 137, 158–164

Morin, F., Coderre, A. (1996): Experiences with "in-house" expert systems. In: CPPA Control Systems '96 Preprints, Forum on Industry. Technical Section CPPA, Montreal, pp. 221–228. (Also Przegl. Papier 53(4), 231–235)

Murtovaara, S., Juuso, E.K., Sutinen, R. (1996): Fuzzy logic edge detection algorithm. In: 1996 IWISP Proceedings. Elsevier Science, Manchester, U.K., p.423–427

Murtovaara, S., Juuso, E., Sutinen, R.; Leiviskä, K. (1998): Modelling of pulp cooking characteristics by using fuzzy logic and linguistic equation models. In: Proceedings of TOOLMET '98 Symposium – Tool Environments and Development Methods for Intelligent Systems. University of Oulu, Oulu, Finland, pp. 90–96

Murtovaara, S., Sutinen, R., Leiviskä, K. (1998): Neural networks modelling of pulp digester. In: Dourado, A. (ed.): Proceedings of CONTROLO'98, 3rd Portuguese Conference on Automatic Control. APCA, Portugal, pp. 627 - 630

Musavi, M.T., Smith, G., Li, J.X., Domnisoru, C. (1999): Neural network for modelling paper optical properties. In: Dagli, C. (ed.): Intelligent Engineering Systems through Artificial Neural Networks vol. 9. ASME Press, 1177–1182

Myllyneva, J., Kortelainen, J., Sutinen, R., Leiviskä, K. (1999): Fuzzy Quality Control of TMP Refiner Line. In: 6th International Conference on New Available Technologies. SPCI, Stockholm, pp. 280-283

Nakatani, H. (1995): Optimized combustion control system for chemical recovery boilers. In: 28° Congresso Anual de Celulose e Papel. ABTCP, Sao Paolo, pp. 983–992

Nault, G., Maltais, D. (1997): Application of an on-line expert system to optimize dryers and press section at Domtar. Pulp Pap. Canada 98(12), 111–113

Nelson, D. (1997): Energy management optimizations using multivariable fuzzy control. In: Engineering and Papermakers: Forming Bonds for Better Papermaking Conference. Tappi Press, Atlanta, 319–322

Nilsson, L., Langkjaer, T. F. (1996): High level control of a causticizing process. In: Zimmermann,H.-J. (ed.): Proceedings of EUFIT'96 – Fourth European Congress on Intelligent Techniques and Soft Computing. Verlag Mainz, Aachen, pp. 1901–1905

Nilsson, L., Dobel, M. (1998): High-level control of the lime kiln and the causticizing process. In: 1998 International Chemical Recovery Conference Proceedings. TAPPI Press, Atlanta, pp. 537–546

Nyborg, L., Crowell, B. (1997): Computer modelling of paper and paperboard for design of expert systems. In: 1997 Technical Conference, Pacific Paper Expo Proceedings, pp. 97–100

Ollila, M. (1997): Online Kappa measurement and control. Pap. S. Afr. 17(3), 19–20, 22–23, 26–28, 30, 32, 34

Ostergaard, J. -J. (1993): Fuzzy control of cement kilns a retrospective summary. In: Zimmermann, H.-J. (ed.): Proceedings of EUFIT'93 – First European Congress on Intelligent Techniques and Soft Computing. Verlag Mainz, Aachen, pp. 552-562

Oyama, Y., Matumoto, T., Ogino, K. (1997): Pattern-based fault classification system for the web-inspection process using a neural network. In: 1997 Japan TAPPI Annual Meeting Proceedings. Japan TAPPI, pp. 575–580. (Abstract in ABIPST 68(10), 1120)

Ozaki, N., Yamazu, S., Tateoka, K. (1994): Recovery boiler intelligent control. In: Control Systems Proceedings. SPCI, Stockholm, pp. 307–312

Powell R.E. (1996): Software for process analysis. In: 1996 TAPPI Engineering Conference Proceedings. Tappi Press, Atlanta, pp. 617–620

Pringle, J.B. (1992): Application of automated vibration signature analysis – concerns and benefits. In: Proceedings Pulp & Paper Maintenance '92, Paper no. 22, 16 p

Pulkkinen, M., Saastamoinen, M., Skyttä, M. (1997): Neuronet and fuzzy logic control oxygen-delignification. Pap. Puu 136(3), 152–154. (Also Pulp Pap. Eur. 2(10), 30.32)

Ribeiro, B., Dourado, A., and Costa, E. (1993): Lime kiln process identification and control: A neural network approach. In: International Conference on Artificial Neural Networks and Genetic Algorithms Proceedings. Springer, Berlin Heidelberg New York, pp. 117

Rudd, J. B. (1991): Using a neural network system for advanced process control. Tappi J. 74(10): 153–159

Rudd, J. (1995): Prediction and control of paper machine parameters using neural network models. In: 28° Congresso Annual de Celulose e Papel. ABTCP, Sao Paolo, pp. 281–286

Samad, T., Gorinevsky, D., Heaven, M., Green, A., Su, T. (1998): Neural network technology for tuning process control systems. In: 1998 Conference on Control Systems Proceedings. Finnish Society of Automation, Helsinki, pp. 195–202

Sasaki, T. (1996): Position correspondence identification algorithm in basis weight profile control. In: XIV IMEKO World Congress Proceedings: New measurements – Challenges and Visions. Finnish Society of Automation, Helsinki, Volume XB, pp. 139–144

Scharcanski, J., Dodson, C.T.J. (1995): Simulating paper forming. In: 1995 International Paper Physics Conference Proceedings. CPPA Technical Section, Montreal, pp. 87–89

Smith, D. B., Edwards, L. L., and Damon, R. A. (1991): Development and application of a real-time recovery boiler expert system. Tappi J. 74(11): 93–96

Sotelo, Junior J., Porchia, L.F.S., Plaut, R.L. (1996): Press simulator used in the development of expert diagnostic system for predictive maintenance. Papel 57(8), 36, 62–64

Stapley, C., Butner, R., Kangas, M., Broeren, L., Smith, B. (1995): Tool for strategic analysis. In: 28° Congresso Annual de Celulose e Papel. ABTCP, Sao Paolo, pp. 331–341

Suojärvi M. (1997): Neural network for estimation of quality. In: XIV IMEKO World Congress Proceedings: New measurements – Challenges and Visions. Finnish Society of Automation, Helsinki, 5 p

Wilks, R. (1997): Production rate optimization, chemiwasher expert system. In: 51st Appita Annual General Conference Proceedings. Appita, Carlton, Victoria, Australia, pp. 535–540

Woollard, R., Jury, J., Clark, C. (1996): Expert system advises operators on Process-Control Conditions. Tappi J., 79(4), 137–141

Xia, Q., Rao, M., Farzadeh, H., Danielson, K., Henriksson, C., Olofsson, J. (1993): Integrated intelligent control system for peroxide bleaching processes. In: IEEE Conference on Control Applications Proceedings. IEEE, Piscataway, NJ, pp. 593-598

Yeager, B. (1997): Fuzzy controls yield clear results at Finnish pulp mill. Pulp Pap. Int. 58(3), 58, 60–61

Zhang, D., Sun, Q., Wadsworth, L. (1997): Model for evaluation of quality of nonwovens and polymeric composites. In: Nonwovens Conference Proceedings. Tappi Press, Atlanta, pp. 223–228

Zhu, J.B., Xia, Q.J., Rao, M., Zurcher J. (1997): Neural Network Modelling Pulp Processes. Pulp Pap. Canada 98(9), 35–38

Enhancing Maintainability of Knowledge-Based Soft Computing Systems

Tapio Rauma
VTT Electronics/Intelligent Systems Group, P.O.Box 1100, FIN-90571 Oulu, Finland

1 Introduction

When a knowledge-based system is under development, the main interest is pointed to the information that is installed in the system, which means how to elicitate the information from domain experts and from other knowledge sources, and how to be sure about the reliability of the information. Fewer interests have appeared towards understanding the whole process of developing knowledge-based systems. In practice, this means combining knowledge engineering and software engineering tasks to build up a development process for knowledge-based systems.

Traditionally, the software development process includes specification, design, coding, testing, and maintenance (see e.g. Boehm 76; Fairley 85) and in knowledge engineering the development process is divided into knowledge acquisition and implementation. In general, we must approve that any development process assures world-class results. The idea of this paper is to present methods to avoid pitfalls of applying poor development practices (if not at all). It is not unusual that the developed system, for example fuzzy or some other rule-based system, produces extremely good results, but it cannot be maintained or further developed because of the total lack of the logical structure, parameterisation, and software documentation. Using such a system is not rational in the long run.

The need for high quality software has always been clear. Considering intelligent systems, the need may be too easy to forget, because the nature of Soft Computing methods (fuzzy logic, neural networks, genetic algorithms) does not guide to use the careful design and development of software. Typical to knowledge-based systems is that in the requirements analysis phase only a slight idea of forthcoming information needs exist. The actual needs appear later, typically during the test period of the system. This makes it difficult to predefine the software architecture, overall demands and test results the system, which are essential when software is developed.

The essential question in developing knowledge-based systems is how to combine changing knowledge, unclear goals, and demands that change during the development work. Additionally, the final result should be a high quality software product, and the system should be easily maintained later. In this paper we will present possibilities to utilise incremental development and modular software to tackle possible problems in the quality of the final software. Two different applications are presented, where incremental development is used. The results show that poor quality can be avoided by the careful development of knowledge-based systems.

2 On the Development of Knowledge-Based Systems

The starting point of the development of the knowledge-based system is the need to represent knowledge in such a form that it can be automatically processed, by a computer. This need may arise from different reasons:

- Only few experts are available: with a proper expert system their knowledge can be utilised by other personnel, or automatically.
- The amount of the knowledge is very large and complex, which may increase the number of errors in human decision making. When the knowledge is installed to an expert system, random errors in decision making may be avoided.
- Although the amount of the knowledge is small, decisions have to be done frequently, which means that the expert is needed all the time. After taking an automatic system into use, the expert can handle other perhaps more challenging tasks.
- Malfunctions due to human errors may cause serious damages, economical losses, or environmental problems.

When we analyse the needs, the first two statements are typical to the systems used in the fault diagnosis area, and the last two are typical in the control systems area.

In general, the main stages of the development of the knowledge-based system are knowledge acquisition and system implementation, which consist of several subtasks (Fig. 1). Some of the tasks can even be considered as a branch of science, for example the interaction between human experts at different levels of knowledge and the interviewer is regularly discussed.

The different kinds of information obtained from varying human experts have been studied widely.

2.1 Knowledge Acquisition

Johannsen and Alty (1991) divide the knowledge acquisition further into knowledge elicitation and machine induction (Fig. 1). Generally, knowledge elicitation is interactive and machine induction is more or less automatic. Lehto et al. (1992)

divide machine induction into interactive (semi-automatic) and automated (machine learning) tasks.

Knowledge elicitation is normally carried out interactively, interviewing domain experts or using other supporting techniques, e.g. questionnaires, walk-throughs, and observations. The purpose is to transfer knowledge from the domain expert to a knowledge engineer, and in this way, make it possible to implement the knowledge-based system.

A typical example of interactive machine induction is knowledge discovery in databases (KDD) (Fayyad et al. 1996). The main purpose of KDD is to find and record automatically detected phenomena for a human expert to analyse whether the phenomena are important or not. The semi-automatic machine induction turns into an automatic one, if requisites for the important phenomena are known and they can be swept automatically.

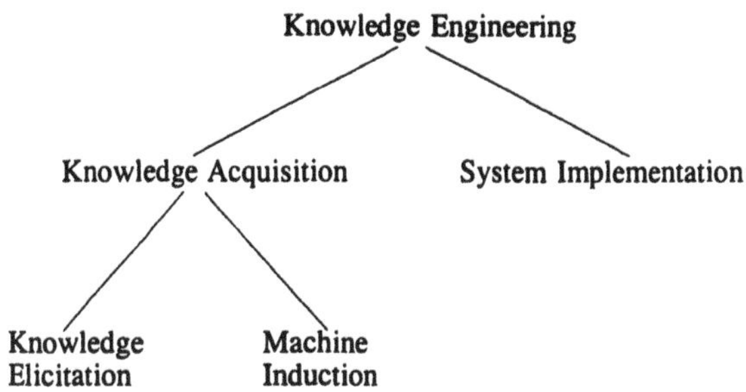

Fig. 1. Relationships between development tasks in knowledge-based systems (Johannsen and Alty 1991).

In the knowledge elicitation phase, there are problem areas that should be taken into account to avoid failures (Johannsen and Alty 1991; Chignell and Peterson 1988; Rouse et al. 1989):

- Co-operating with too few experts, or with experts of single level
- Conflicts: several experts can not agree with their information
- Collecting knowledge only from the most convenient sources
- Limited human understanding
- Difficulties in handling the knowledge elicitation process itself, for example biases. The amount of knowledge of the knowledge engineer increases during the elicitation, which sets domain experts on a different starting-point in interacting with the knowledge engineer. The knowledge engineer must avoid biasing knowledge according to the already interviewed experts
- From a domain expert's view, knowledge transferral may be difficult, because it may be the first time he/she has to explain the knowledge in a formal way, or in a way that makes it possible to formalise the knowledge.

Additionally to these factors, there are practical problems that can be divided into three groups: the capability of domain experts (1), the skills of an expert acquiring the information (2), and the acquisition process itself (3). These factors can be, for example (Johannsen 1990; Gaines 1987):

- Expertise can be fortuitous, which means that the expert does not know all the affecting details. (1)
- An expert can only present information, not give any critique to information given by other experts. (1)
- The knowledge engineer can not understand the language of the domain expert. (2)
- Expertise may not be linguistically expressible (the domain expert is unable to express the information). (1) and (3)
- Expertise may not be applicable. Due to the complex or unique nature the information is impossible to utilise in the developed system. Perhaps the most common case is when an expert can easily provide important information, but it is useless because the characteristics of the information can not be measured. (1) and (3)
- Expressed expertise is irrelevant or presented at an unfavourable stage in the knowledge acquisition process. (2) and (3)
- Expressed expertise is incorrect. (1)
- Expressed expertise is incomplete. (1) and (3)

Typical to knowledge acquisition is that there seems to exist much knowledge, but formalising and putting different granules of information into right levels of importance is difficult. During knowledge acquisition experts of different domain areas have to try to put their information into a form that can be utilised as a computer program. As Buchanan et al. (1983) initiates, knowledge acquisition can be seen as a transfer of problem-solving expertise from some knowledge source to a program.

2.2 The VVT and Implementation

This paper concentrates on systems that are used on-line, parallel with dynamic industrial processes. These kinds of systems and tasks are, for example, operator supportive systems, control systems, fault diagnosis systems, consequence prediction, and procedural support (Johannsen 1990).

Unlike off-line systems, which are commonly used in analysing measured information, on-line systems can not usually be tested before installing them into the target environment. Additionally, on-line systems are normally critical when we examine possible risks of applying a new system into the target environment. That is why the whole development process as well as individual tasks in different stages of the development process must be determined carefully.

The VVT (Verification, Validation, Testing) process is carried out to ensure the quality of the development work during and after the knowledge acquisition stage. The goal of the VVT process is to prevent undesired malfunctions and problems caused by incomplete knowledge transfer between domain experts and knowledge engineers. When malfunctions are found early in the process, fewer errors will appear later on during the on-line testing of the knowledge-based system.

The purpose of verification is to clarify, whether the system has been technically constructed correctly. Typical tasks of verification are to analyse the completeness of the knowledge base and other essential characteristics of the system, such as the size and the complexity of the system, if the complexity can be measured. (Lopez et al. 1990)

Validation concentrates on analysing the behaviour of the system, in other words whether or not the system is acting like it should. Validation is seen as a more difficult task than verification, because generic validation methods do not exist. The most common approach is to utilise domain experts to analyse the developed system. This works fairly well for small knowledge bases, but as Golshani (1990) maintains:

> "...for complex problems, which may require a knowledge base containing tens of thousands of rules, we will need a lot more than the confidence of an expert (or a group of experts) before we can use the system for anything serious."

Testing means the evaluation of the developed knowledge-based system in the target environment. In an ideal case, malfunctions have been already avoided in the verification and validation phases, and when tested no severe, knowledge engineering related errors are found. But it is probable that the operation of all features of the developed system can not be ensured beforehand, which means that errors are found and corrected after installing the system in the target environment.

2.3 Special Features of Fuzzy Knowledge-Based Systems

Several knowledge sources to exploit in developing fuzzy systems can be found in the literature. The typical ones are, for example, expert knowledge, mathematical process models, fuzzy process models, learning, and operators' control actions (Takagi and Sugeno 1983; Lee 1990). Depending on the case, other knowledge sources can be included in the list, e.g. the documentation of an existing control system, and records of earlier attempts to solve the problem.

When the knowledge-based fuzzy systems are analysed, it may be noticed that risks are involved in the development work. Especially if the incremental development process is used; the following items should be taken into consideration:

- Multiple architectures and methods exist that can be exploited in developing a fuzzy system. An incremental development process must be taken into account when choosing the architecture to ensure the possibilities to add new parts to the existing system.

- Fuzzy logic imitates the thinking of a human (in some amount). This property should remain as far as possible during the development work. The iterative knowledge transfer between the tested system and knowledge acquisition necessitates this.
- Lack of knowledge. An incremental development process helps to increase the amount of knowledge during the development cycles. However, some pieces of information should be clear and exploitable when the development is started.
- Due to typical cross-references in fuzzy rule bases, the maintenance of a fuzzy system is more difficult than maintaining simpler rule-based system. There is a risk that incremental development makes the system complex to maintain.

When we examine the incremental development process together with the problems presented above, we can formulate the problems that exist in developing a knowledge-based fuzzy system using an incremental development process. Based on the problems that exist in knowledge acquisition and in choosing the architecture and methods, the key problems are

> How to avoid the risk that the development process turns into a trial-and-error procedure?
> How to keep the (software) product easy to maintain?

3 Development Processes of Knowledge-Based Systems

One important factor in the development of knowledge-based systems is to choose the most suitable practices to carry out the development. As we earlier presented, knowledge engineering is normally divided into two main tasks: knowledge acquisition and system implementation. These two tasks can not be kept separate as they usually are presented, which causes difficulties in the development process.

In the next section we first take a look at the s. c. straightforward approach, which may be useful if the two tasks can be kept separate. The other methods to be discussed are based on an assumption that the results of the knowledge acquisition stage are generally not complete, and the knowledge must be augmented during the system implementation.

3.1 A Straightforward Approach

The most common way to express the straightforward approach is to exploit the simple waterfall model that is commonly used in software engineering (Boehm 1976; Fairley 1985, Fig. 2). When this model is applied to knowledge engineering, the first two tasks can be carried out in the knowledge acquisition stage and the rest in the system implementation stage. Inside the main tasks we can name several subtasks, for example designing, coding, installing, and testing, which can be included in the implementation stage. Characteristic to this approach is that it is assumed that the previous tasks have been executed so well that there is never the need to turn back in the development process.

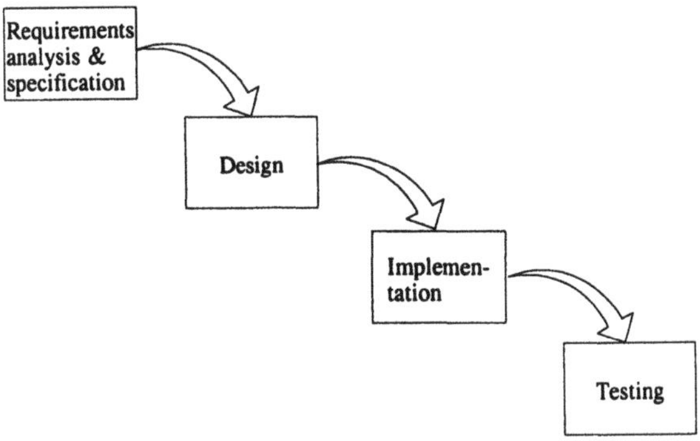

Fig. 2. The waterfall model - a simplified version.

As Gomaa (1993) states, from the viewpoint of software engineering, the most significant problems with the waterfall model are

- Software requirements are not properly tested until a working system is available to demonstrate to the end users. It has been shown that the errors in the requirements specification are the last to be detected, and therefore they are the most costly to correct (also Boehm 1976).
- The developed system will be available late in the development process, which means that some major design or performance problems may arise late during the development process.

When we put the waterfall model into practice in knowledge-based systems development, the above mentioned statements may cause problems. Based on our experience, especially in the development of the knowledge-based fuzzy systems the acquired information sets the limits to exploit the waterfall model. It is typical that expert knowledge becomes more accurate after the expert has real-world test results of the developed system.

Aforementioned facts show clearly, why the waterfall model type is not practical in the development of the knowledge-based systems. The waterfall model may be used only if the problem to be solved is simple enough, so the needs and influencing characteristics are easily understood.

3.2 An Incremental Development Approach

The origin of the incremental development approach is in the research of Boehm (1988). The spiral model of Boehm was developed for software engineering, but later it has been applied to, for example, expert systems development (O'Keefe and Lee 1990) and fuzzy systems development (Isomursu 1995; Isomursu et al. 1995).

The basic idea of the spiral model is presented in Fig. 3. Different phases of the development process are cycled iteratively during the development process until the requirements are fulfilled. With this approach, the achieved results can be taken into account when the consequent actions are planned and analysed.

Considering the suitability of the iterative approach to knowledge-based systems development, it is important that the approach makes it possible to proceed in small steps, until the final solution is achieved. With small steps the reliability of the development work in all phases of the development process can be more assured than in the straightforward development process. Because of small steps, the risks of the development work are reduced into smaller aggregates that are directly related to the amount of the development work per round.

In this paper, we apply the incremental development process for the following reasons (Rauma 1999):

- The straightforward approach is often unsuitable for the applications of control, diagnosis, and expert systems due to the complexity of even conventional examples. For example, the goals of the system can not be set before testing the first versions of the developed system.
- The incremental development process supports using a modular system architecture. Modular architecture has proved to be effective especially in control applications.
- The incremental development process makes it possible to exploit all the information that is acquired during the development of the system. In the straightforward approach, only a priori information is available.

On the other hand, the following disadvantages can be stated:

- The development work can be started with a too little amount of information. This can make the early efforts of the development work frustrating.
- If the problem and its possible solutions are predetermined and pre-evaluated poorly, the development process becomes very easily similar "a trial-and-error" strategy.
- The more rounds that are needed in the incremental development, the more risks appear concerning the quality of the development work. This especially concerns documentation, test results, system architecture, and software implementation (if produced).
- It may be difficult to estimate the efforts needed during the development.

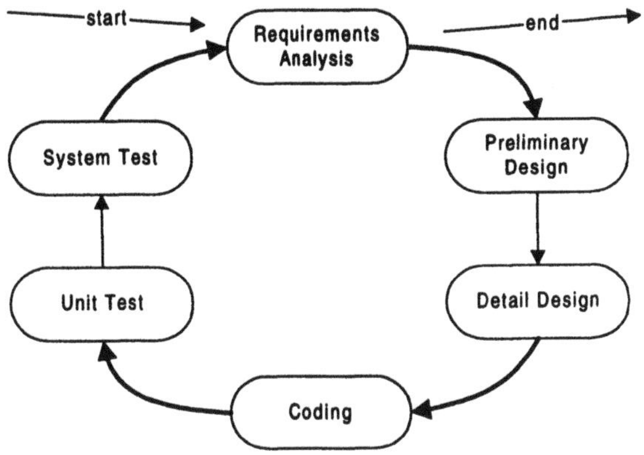

Fig. 3. The idea of Boehm's spiral model.

- The process is based on achieved results and possibilities to enhance the results. This means that it is more comfortable to see the possibilities to enhance the operation of the system than compare the results to the requirements.

4 Putting the Methods into Practice

4.1 Modularising Systems

This paper is based on the research conducted during the last years, and two levels have been defined to apply the modular architecture:

- A large total control system can be built using distributed modules with the shared control strategy.
- A large single control problem can be solved using partial development and modular architecture.

Common for both alternatives is that they could well be solved by developing only one single software module, where all needed information and decision making is embedded. The first case is actually not highly abnormal compared to modern control engineering; the main point of this is the control strategy.

In modern control systems, the controllers are built from simple controlling units and the total controller can be seen as a net of simple controlling units. In such case the control strategy that was chosen when the development work started, is embedded in the network of controlling units. In our perspective, distributed knowledge-based modules are more intelligent than simple controlling units, and can therefore utilise the control strategy independently, without the "controlling network" that is the alternative way.

The second alternative way to modularise for example control systems is highly affected by incremental development process. The method is presented in (Rauma 1999) and is successfully applied in several applications. In the next chapters we will present examples of both modularising methods. The both applications are developed in Outokumpu Zinc's Plant in Kokkola, Finland, and the developed systems have been in daily use for several years.

4.2 Controlling a Gas Purification Process

An example of modularising a control system of the type 1 is a control system for a two-stage gas purification process for sulphur dioxide. In the first stage the gas is cooled (with sulphuric acid) to about 200°C. In the second – more important – stage the gas is then purified with sulphuric acid (at around 40%–70% H_2SO_4). The sulphuric acid becomes hot – from 40°C to 60°C – in the process and it has to be cooled with cold water in a special acid cooling unit. Water used in the acid cooling unit is cooled in five water coolers in the open air. The capacity of the acid cooling unit as well as the water cooling system is controllable.

In Outokumpu Zinc's factory there are two identical purification processes. The processes are not independent since the cooling water used in the acid cooling unit is common. Control actions in the water cooling system affect both purification processes. This makes it essential to choose a reliable control strategy before developing any control system. The problem comes from the structure of the process. Because cooling water used in the sulphur acid cooling units is common for both processes, change in the cooling tower's capacity affects both processes' temperatures. This problem appears especially when processes are (for some reason, e.g. disturbances) in the different state. The problem prevents the sole use of water coolers as a main controlling device. Some of the control must in these cases be done in sulphuric acid's amount, which is not – due to constant environment requirements – desirable.

To achieve a good purification result, the amount of sulphuric acid used should be as constant as possible (usually at around 100 m^3/h) and the temperature of the gas after the process should be even more constant (deviation around ±0.5°C). Acid cooling unit and the water coolers control the temperature of the sulphuric acid. Because the amount of sulphuric acid should be as constant as possible, the main control should be done using the acid cooling unit and the water cooler's capacity.

Dead-times of the process are appreciable. For example, the delay between a change in sulphuric acid volume and temperature change in the gas is several minutes. Additionally, a change in the water coolers' capacity is detected in other parameter values after ten to fifteen minutes.

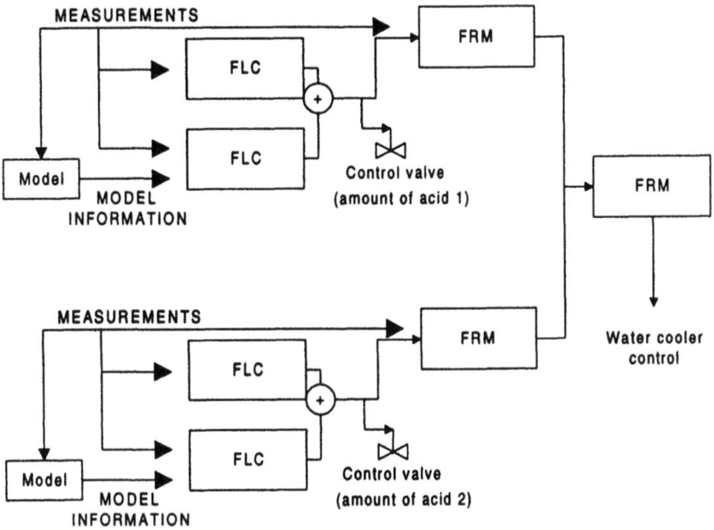

Fig. 4. The structure of the final control system of the gas purification process.

This application is here referred as the type 1 (see Chapter 4.1): *A large total control system can be built using distributed modules with shared control strategy.* The structure of the final control system is presented in Fig. 4. As mentioned earlier, in this kind of applications the control strategy is embedded into single controlling units.

More detailed presentation of the control system can be found in (Rauma 1999). However, the main points of the control system are:

- The modular architecture together with incremental development made the structure of the control system clear, which made it possible to understand the behaviour of the system by means of understandable software documentation.
- The incremental development gave confidence to the development work because preliminary results could be achieved before the final system was installed and tested in the real environment.
- The system has been in continuous use for almost five years, which can be considered as a quite long period, from the point of view of the knowledge-based systems.

4.3 A Temperature Controller

An example of modularising a control system of the type 2 is a temperature controller of a zinc roaster furnace. Roasting zinc concentrates produces heat (exothermic reactions). The concentrates (solids) are sulphides and react with oxygen at high temperatures, forming calcine (solid) and sulphur dioxide gas. At the Kokkola Zinc plant of Outokumpu, there are two almost identical continuous roaster furnaces with the total capacity to feed about 355 000 tons of zinc concentrates (Zn 53 %) per year. The furnaces are 17 m high and have a basal area of 72 m^2 each. As well as sulphur dioxide, the plant also produces steam as a by-product.

The temperature of the furnace is the most essential factor when the quality of the product, which is calcine, is controlled. In general, the temperature of furnaces must be over 900°C in order to keep the process going, and below 980°C to avoid damage to machinery and possible sintering effects. Used maximum temperature changes (usually between 930°C and 960°C) depend on the amount of different harmful chemical elements, such as lead (Pb) and copper (Cu). In general, the temperature should be kept as high as possible to make the level of produced zinc calcine high, but at the same time the personnel have to be careful not to exceed the maximum temperature.

The temperature of the roaster furnace is controlled by changing the volume of concentrate-feeding devices (which means changing the volume). Before developing the control system the process was controlled manually. Operators made control actions typically once per ten minutes or more frequently. The deviation of the temperature was usually around ±10°C during the normal use. Some efforts had been put on installing and tuning a normal PI-controller to control the temperature, but without success. Remarkable delays and disturbances make the problem too difficult to handle with simple systems.

In this case the incremental development of the control system comprised three parts:

- Developing and installing a basic controller
- Developing and installing a corrective module into the control system
- Developing and installing a meta-rule module into the control system

The structure of the final control system is presented in Fig. 5. Compared to the previously presented application (Chapter 4.2), the key point is that the same single control problem is solved by parallel controlling modules instead of implementing a certain control strategy with using several independent controlling units.

At first, a normal fuzzy logic controller with nine rules (three membership functions for each variable) was constructed and installed in the Alcont automation system using the Pascal language and normal logic-based programming methods (typical to the Alcont system). Parameters of the controller were tuned manually during a period of several days. The control results, achieved by using the basic controller, were satisfactory. However, the overall behaviour of the control system was not always acceptable. For example, the controller made changes in feeding in cases where changes were not needed, which caused unnecessary effects to the process.

After testing the basic controller, it was found that despite the unreliable behaviour of the residual oxygen measurement, it would be advantageous to utilise the level of residual oxygen in the control system. Changes in the residual oxygen seemed to predict the behaviour of the temperature reasonably well. Therefore it seemed possible that utilising the changes in residual oxygen would make the control system more reliable. This is why we decided to use an additional fuzzy module (a corrective module) to take residual oxygen into account. This fuzzy module used the change in the temperature in the furnace (three membership functions) and the change in the level of residual oxygen (five membership functions) as input variables. The output of the module is the required change in the capacity of concentrate feeding devices, which is added to the output of the basic controller.

After achieving satisfactory results another effort was made to make the recovery from a faulty situation as easy as possible. When recovering to the normal temperature it is essential that it occurs as fast as possible, but not by allowing the temperature to exceed the set point. The meta-rule method presented in (Isomursu et al. 94) was used to control the recovery of the temperature. In this case the meta-rule approach is used to keep the temperature increase at the suitable level. In practice, the temperature is allowed to increase rapidly if the offset is large, and the permitted increase in the temperature is reduced smoothly, using a meta-rule module, when approaching the set point. Basically, a PI-type controller does this well in normal cases, but it was discovered that with this approach the recovery is quicker and more efficient. With a normal PI controller, "fast" would have meant more overshoot, but with a meta-rule based fuzzy system the overshoot could be avoided.

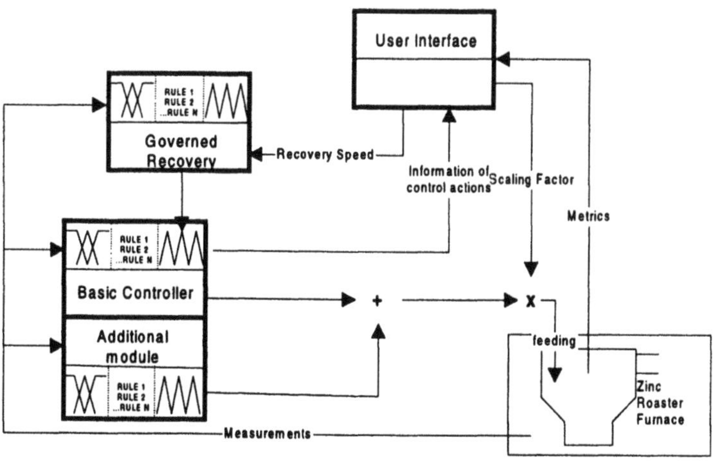

Fig. 5. The structure of the installed control system.

A detailed presentation of the control system can be found in (Rauma et. al. 2000), where also control and other results are presented. The main points of the control system are:

A more stable process

- Compared to the previously used manual control, the fuzzy control system has reduced the general deviation of the temperature from 10°C to 3°C (95% of the time).
- The improved temperature control made it possible to also use concentrates of low quality.

Capacity of the roaster has been increased

- Because of better temperature control, furnaces operate more accurately near the set point; in other words: near the theoretical maximum production level. No exact information can be presented, because other (mechanical) improvements in the process, which affect the capacity, have been carried out during these years.

Less sintering

- The set point of the temperature is not usually exceeded, which prevents dirt. This also increases capacity.

Less effort to control the temperature

- When the concentrate feeding was controlled manually, operators had to monitor the behaviour of the temperature continuously. The need for continuous monitoring almost disappeared with the automatic control system.

5 Conclusions

One apparent risk in developing knowledge-based systems for real-time use is the poor quality of software. The development of the knowledge-based systems is typically not the straightforward software development, which hinders using strict specifications and guidelines. Software is usually developed step by step, and new needed features are specified after using the test results of the incomplete system. The development of the software should be carefully organised to avoid losing the quality. If the quality of the software is poor, it makes the continuing development difficult and may make it difficult to maintain the system later.

In this paper different possible development practices, development processes, and structures of knowledge-based systems were analysed. The main motivation is to formalise the development of the fuzzy logic based systems, but the results of the studies can be applied also when other techniques are used. Compared to general software engineering, the nature of the knowledge-based systems, and their development, sets unique requirements. It is more than usual that the level of knowledge increases during the development work, which means that planning and design actions must be arranged parallel with the implementation and testing phases.

The incremental development process together with the modular system architecture is in this paper seen to be the most efficient, when the fuzzy knowledge-based systems are developed. When modularising is successfully done, the development can be carried out via developing and installing new modules to the existing system without changing older ones at all. The two approaches presented are based on distributing the total problem into smaller ones, with taking the nature of the problem into account. Presented applications gave good experiences in both cases.

Presented examples (applications) represent two different approaches to utilise the incremental development and modular architecture. Both are difficult control problems with unique demands. The examples show how to implement modular architecture in problems of different kind of nature. In both cases the control systems have been in continuous use for several years, which verifies the quality of the development work.

One purpose of this article is to bring out a way to formalise the development of fuzzy systems. In knowledge-based systems' area it is essential to avoid the trial-and-error procedure during the development work, because it usually means the more complex development process and, because of that, a high risk for poor results. One important way to make knowledge-based systems more reliable is formalising the development process as well as other development practices. This may make knowledge-based systems more attractive even in the most demanding applications.

References

Boehm, B.W. (1988): A spiral model of software development and enhancement. IEEE Computer 21(5), 61–72

Chignell, M.H., Peterson, J.G. (1988): Strategic issues in knowledge engineering. Human Factors 30, 381–394

Fairley, R. (1985): Software engineering concepts. McGraw-Hill, New York

Fayyad, U.M., Piatetsky-Shapiro, G., Smyth, P. & Uthurusamy, R. (eds.) (1996): Advances in knowledge discovery and data mining. AAAI (American Association for Artificial Intelligence) Press, Menlo Park, California

Gaines, B.R. (1987): An overview of knowledge acquisition and transfer. Int. J. Man-Machine Studies 24(4), 453 – 472

Golshani, F. (1990): Rule-based expert systems. In: Adeli, H. (ed.): Knowledge Engineering. Vol. I. Fundamentals. McGraw-Hill Publishing, New York, pp. 28 – 51

Gomaa, H. (1993): Software design methods for concurrent and real-time systems. Addison-Wesley Publishing, Menlo Park, CA

Isomursu, P. (1995): A software engineering approach to the development of fuzzy control systems. VTT Publications 230, Technical Research Centre of Finland, Espoo

Isomursu, P., Rauma, T., Kemppainen, S. (1995): Applying a model of fuzzy control software development process in practice. In: Proceedings of the Eight International Conference on Industrial & Engineering Applications of Artificial Intelligence and Expert Systems (IEA/AIE'95). Gordon and Breach Science Publishers, Melbourne, Australia, pp. 391 – 400

Johannsen, G. (1990): Towards a new quality of automation in complex man-machine systems. In: Proceedings of IFAC 11th World Congress on Automatic Control. Pergamon Press, Oxford, vol. 10, pp. 175 – 181

Johannsen, G., Alty, J. L. (1991): Knowledge engineering for industrial expert systems. Automatica 27(1), 97 – 114

Lehto, M.R., Boose, J. Sharit, J., Salvendy, G. (1992): Knowledge acquisition. In Salvendy, G. (ed.): Handbook of Industrial Engineering. 2nd ed. John Wiley & Sons, New York, pp. 1495–1545

Lopez, B., Meseguer, P., Plaza, E. (1990): Knowledge based systems validation: A state of the art. AICOM 3(2).

O'Keefe, R.M., Lee, S. (1990): An integrative model of expert system verification and validation. Expert Systems with Applications 1, 231–236.

Rauma, T. (1999): Fuzzy modeling for industrial systems. VTT Publications 382, Technical Research Centre of Finland, Espoo

Rauma, T., Nyberg, J., Herronen, J. (2000): Fuzzy control of furnace temperature in a zinc roaster. In: Jämsä-Jounela, S-L., Vapaavuori, E. (eds.): Proceedings of the IFAC (International Federation of Automatic Control) Workshop on Future Trends in Automation in Mineral and Metal Processing. Finnish Society of Automation, Helsinki, Finland, 161–165

Rouse, W.B., Hammer, J.M., Lewis, C.M. (1989): On capturing human skills and knowledge: Algorithmic approaches to model identification. IEEE Trans. on Systems, Man, and Cybernetics 19, 558–573

Takagi, T., Sugeno, M. (1983): Derivation of fuzzy control rules from human operator's control actions. In: Proceedings of the IFAC Symposium on Fuzzy Information, Knowledge Representation and Decision Analysis, Marseilles, France, pp. 55–60

Part II

Mineral and Metal Industries

Fuzzy Logic Control of a Sintering Plant

Pekka Myllymäki[1], Jussi Poutiainen[2]
[1] Rautaruukki Steel, Primary Products, P.O. Box 93, FIN-92101 Raahe, Finland
[2] Filtronic Comtek Finland, Takatie 6, FIN-90440 Kempele, Finland

1 Introduction

Sintering process is an agglomeration process in which fine grained iron concentrates are sintered in high temperature to particles of bigger sizes, which are further used as iron bearing burden in the blast furnace. The sintering process uses a lot of energy, its material flow is very large and maintenance costs are high, so considerable savings potential exists. Attempts to automate it have been done, but really good, simple and working control system has not been developed. A reason for this is the process itself; it is too complicated and multivariable to the traditional control methods.

The development of new control methods, fuzzy logic and neural networks, and the leaping development in the computer technology, has made it possible to control the multivariable processes as sintering. Early control systems were, however, based on measurement technologies difficult to maintain, e.g. IR-cameras. But the idea how to use these new control methods to handle these kinds of processes got support widely. After the 1980s, many variations of the new control systems were developed around the world. A common feature for them all is that they are made for one and only particular sintering plant, not generally applicable for all sintering plants.

Maki et al (1991) report a fuzzy control system for the sintering plant that is based on the measurement of the burning-through point (BTP) of the sintering strand. The burning-through point is considered as a position where the burning zone reaches the bottom of the sintering bed and the mixture layer is completely agglomerated. It is desirable for high production that this point is near to the discharge end of the sintering machine and also a stable temperature distribution across the strand is desirable in aiming to stabilise process and sinter quality. Five series of thermocouples, four thermocouples each, measure the temperature distribution in the wind boxes located under the strand (see Fig. 1). Adjusting the bed height with controllable openings in the feed end controls the temperature distribution in cross-strand direction. An ultrasonic level sensor measures the bed height and a cascade fuzzy controller is responsible for controls.

Nakajima et al (1991) replaced the BTP by burn-rising point (BRP) that is easier to define. In their case it means the point where the temperature reaches 250°C in machine direction. Interpolating between successive series of temperature measurements defines this point. The strand speed controls the BRP.

Arbeithuber et al. (1995) used fuzzy control with the strand speed to keep three successive temperature measurements inside the set limits and, therefore, assuring an optimal temperature profile of the strand. Even though this approach leads to continuously adjusting the strand speed, it maximises the production and improves sinter quality.

Jang et al. (1998) modelled the operators' actions using a multi-layer perceptron network (MLP) and used the network in the strand speed control. Hu and Rose (1997) used a predictive fuzzy controller in the strand speed control.

The system developed in this paper is a combination of previous approaches adapted to the plant in question. The fuzzy strand speed control was developed to manage the burn-through of a sintering bed on a Dwight-Lloyd-type sintering machine. The controller measures the position of a burn-rising point (BRP) along the sintering strand and controls this position by the strand speed. Here, BRP is considered as a position where the waste gas temperature rises over 180°C (Poutiainen 1999). This position is defined in practice as a point in a certain waste gas temperature curve measured from the wind boxes under the strand by recently installed temperature measurement lines. The burn-rising point is used instead of the burning-through point as the BTP is difficult to measure in practice.

The major targets of the project were to decrease the amount of work needed in process control and also to increase the sinter production. At the same time it is possible to fulfil other targets: decrease fuel rate, prolong plant life, stabilise sinter quality and decrease the amount of return fines and dust. All these targets can be reached with the better management of the burning-through point. These improvements do not need big investments, just a few measurement lines, a logic control box and the field bus network. After these low cost investments it is possible to increase the automation level of the sintering plant, especially the part of the burn-through management.

During the test period fuzzy control worked correctly. It controlled the strand speed more accurately and more effectively than manual operation during the comparison period.

2
Sintering Process

Continuous Dwight-Lloyd-type sintering machine is presented in Fig. 1. Three pieces of this kind of sintering machines exist at Rautaruukki Steel Raahe works with the annual total sinter production of about 2,85 million tons. Figure shows also the five new temperature measurements lines located close to the discharge end of the sintering machine.

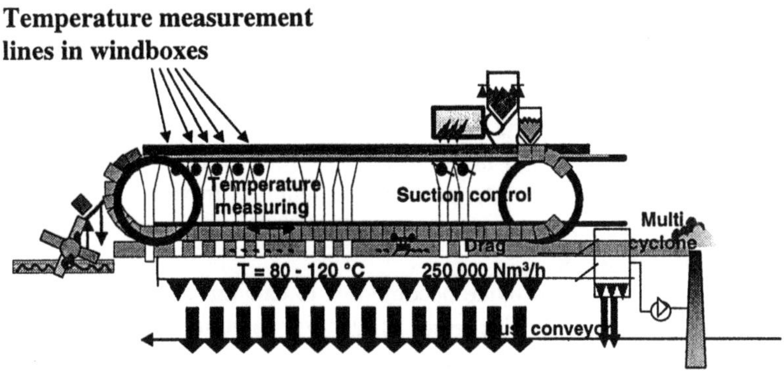

Fig. 1. Sintering machine (Rautaruukki Steel Raahe works).

Mix feeding equipment feed sintering mix containing iron ore concentrates, additives and coke breeze onto a moving hearth called the sintering strand at the right side of the figure. It is build up of many sintering pallets moving on rails. Sintering mix forms a sintering bed, which is sintered on the strand as it proceeds towards the discharge end. Moisture, permeability, grain size distribution, fuel content, chemical and mineralogical composition and stable quality and material flows are the controlled parameters of the mix and reaching these targets is an essential requirement in the sintering mix handling stage. The mix has to be charged onto the strand carefully and the bed must have a good permeability and the surface has to be even. Permeability distribution of the bed across the strand has to be even and the burn-through must happen in the desired position near the discharge end of the strand. Good reducibility, reduction degradation index, cold strength, high temperature properties and desired chemical composition are vital demands for the sinter quality.

A coke gas burner ignites coke breeze in the upper layer of the sintering mix bed under the ignition hood (just above "suction control" in the figure). Burning air flows through the ignited bed from the top towards the bottom of the bed because of the negative pressure inside the sealed wind boxes situated below the strand. Negative pressure is effected by a waste gas blower, which blows waste gas to a chimney. Burning zone (about 1300°C) moves through the bed from the top towards the bottom while the strand continuously moves ahead.

After the burning zone has passed through the bed, which takes about 15–20 minutes depending among other things on permeability of the bed, the sintering bed has been totally agglomerated and sinter is ready to discharge from the strand to a sinter crusher. Strand speed has to be controlled so that the burn-through happens as near the discharge end as possible and position of burn-through must also stay stable. After crushing the sinter flow is screened as hot, cooled in a cooling tower and finally screened as cold before conveying to the blast furnaces. Undersize material from screening is circulated back to the mix.

The sintering bed can be divided to seven zones according to reactions and phenomena occurring in the bed. From the bottom to the top of the bed these zones are:

- The wet zone is formed from the sintering mix and water flowing downwards in the bed. To prevent too high moisture content, burnt lime is added to the mix. This lime is reacting with water producing calcium hydroxide.
- In the drying zone the temperature rises to 100–120°C and water evaporates and drying of the mix starts.
- In the next zone, calcium hydroxide decomposes to calcium oxide that further reacts with carbon dioxide in flue gases producing calcium carbonate. Other hydroxides react according to the same reaction at the temperatures of 400–500°C.
- Carbonates decompose in the calcination zone at the temperature of about 800°C.
- The next zones, burning and sintering zones, are the most important for the product quality. Ferro oxides reduce in this zone because of carbon oxide present in abundance. Molten slag matrix is formed in the sinter.
- In the last zone, oxidisation and cooling zone, ferro oxides is oxidised back again.

In practice, the boundaries between these zones are not clear and their importance varies when going from the feed end towards the discharge end.

3 Burn-Through Management by Fuzzy Speed Control

3.1 Burn-Through Measurement

The wind box temperature measurement system has been installed at the sintering machine 2. There exist 5 probes together in the wind boxes 10–14, one probe in each wind box. Each probe contains 8 thermocouples, so it means 40 thermocouples together. Wind box probes have been tested for 14 to 20 months and there have been no problems with measuring or wearing.

The weighted average of the waste gas temperature is calculated for each wind box. The second order curve is fitted to these weighted wind box measurement values to calculate the location of the burn-rising point (BRP) which is defined as a distance from ignition point to the point where 180°C is attained in the curve.

The BRP is used instead of the burn-through point (BTP) because in Rautaruukki's process the sinter cake is discharged from the strand very near to the discharge end and therefore the BTP is very difficult to measure in practice. While using the BRP instead of the BTP, measurement and control are easier to arrange. Definitions of the BTP and the BRP values are presented in Fig. 2.

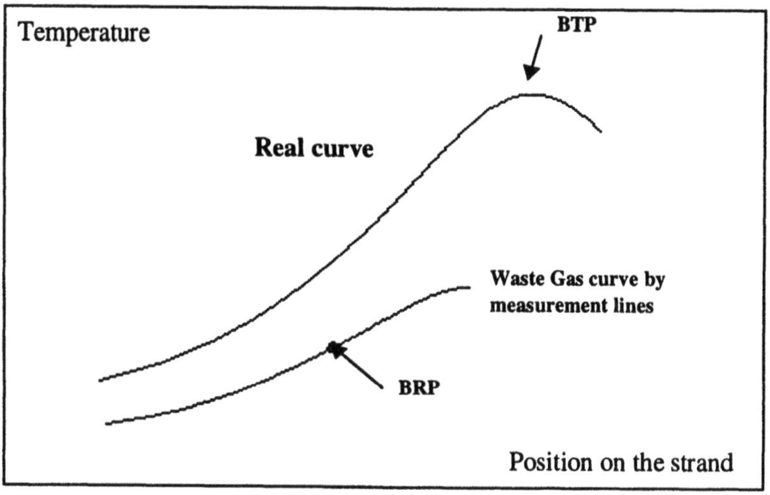

Fig. 2. Determination of burn-through point (BTP) and burn-rising point (BRP) from curves fitted to temperature data measured by the wind box measurements.

3.2 Fuzzy Development Environment

Development environment for data acquisition and fuzzy control has been built and tested during years 1998–2000. The equipment consists of the temperature measurement lines, a programmable logic controller and a PC with Windows NT 4.0 operating system. This development system enables developing and testing of fuzzy logic applications and tuning on-line in the sintering process. Fuzzy system was developed in commercial FuzzyTech program environment (von Altrock 1995).

3.3 Control Strategy

Sintering process is a complex and multivariable system, which includes long delays. Process is also exposed to different kinds of fault situations and periodic fluctuation. Therefore the developed control is a multilevel controller including fast and slow controllers. The fast controller finds out larger limits around the set points and handles unexpected fault situations. The fast controller uses just slightly filtered value of the BRP as input and enables fast response in the case of sudden bigger disturbances.

The slow controller is a fine tuning controller, which tunes the BRP position measured over a long time period (level) to the set point. The slow controller uses a kind of long-term mean value of the BRP as input and keeps the value of the controlled parameter (strand speed) at the right level despite of periodic disturbances.

An expert knowledge has been used to define the optimal control intervals e.g. time between the control actions. It takes about 15 – 20 minutes after the response in the BRP is fully seen after the change in the set point of the strand speed. Therefore it is not possible to change the set point after each control cycle. The control cycle includes measurements, calculations and fuzzy logic control and it takes 40 seconds. However the period between the set point changes in the strand speed can be shorter than 20 minutes, because the response is partly seen earlier from the measurements. The non-linear fuzzy controller handles the time delay management.

There exist also the fuzzy logic controller for waste gas temperature and sinter temperature management (Myllymäki and Poutiainen 2000). The waste gas temperature has a minimum and sinter temperature has a maximum limit, which are not allowed to exceed. If either of these limits are approached, fuzzy logic changes the BRP set point of the BRP fuzzy controller so that the controller slows down the strand speed and the sinter cake is cooled more properly already on the strand. The BRP set point is changed back to its original value after the disturbance is over. The overall control scheme is presented in Fig. 3.

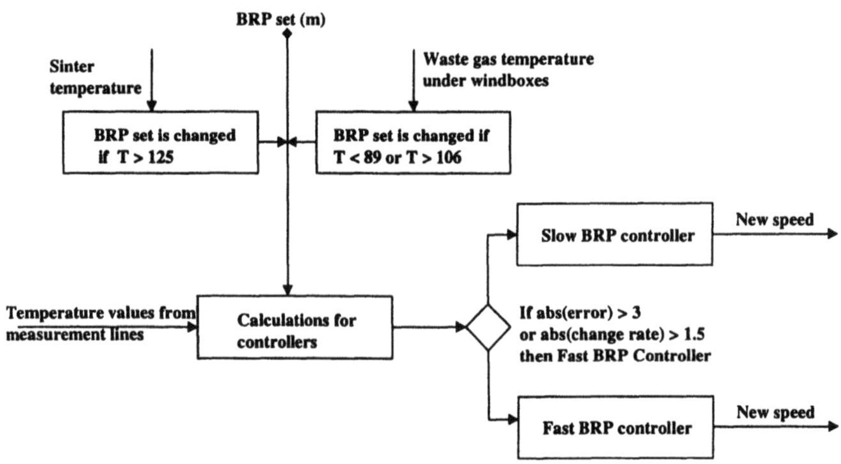

Fig. 3. Fuzzy control for sintering machine burn-through management.

The BRP value seems to correlate very strongly with the average waste gas temperature, which has been the main parameter previously in the strand speed control. It has been shown that typically the BRP value lies between 23–28 m and as a mean value 25,5 m has been used as the set point of the BRP so far. The length of the sintering machines in the Rautaruukki works is 30 meters. The operator can change the BRP set point from the monitor of the basic automation system.

3.4 Fuzzy Controllers

3.4.1 The Fast Controller

The fast controller is used when the process is moving strongly away from its set point, mainly because of larger disturbances. It is not used to compensate for normal periodic disturbances hence its control limits are quite far away from each other. Mainly the fast controller works only in fault situations when it returns the controlled BRP inside its limits and afterwards, the slow controller takes care of the smoother controlling.

The fast controller uses the error and the change of error as input variables. The change in the strand speed is the output variable. The error is calculated from the equation

$$e(k) = BRP_{setpoint} - BRP_{average} \qquad (1)$$

where $BRP_{average}$ is calculated from the last eight burn-rising points, $BRP_{setpoint}$ is the set point for the burn-rising point and k refers to the control interval. The change of error is calculated as

$$\Delta e = e(k) - e(k\text{-}4m), \qquad (2)$$

where e(k-4m) is the four minutes older error.

Input variables are calculated in the every control cycle, in this case after each 40 seconds. When the change in the strand speed is done, the operation of the fast controller is frozen for the next 20 minutes. This is the time required for the change to take place. Fig. 4 shows the fast controller's layout. Input variables are fuzzified using membership functions as shown Figs. 5 and 6. Membership functions for the speed change are in Fig. 7.

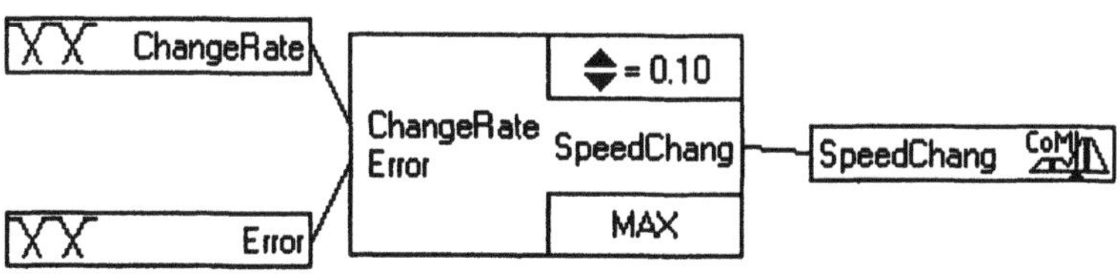

Fig. 4. The layout of the fast controller.

Fig. 5 shows a considerable dead band around the set point. This assures that the fast controller reacts only to bigger disturbances and not to the normal periodic variations. After the error goes beyond its limits, the fast controller reacts strongly and returns the operation inside the allowable region.

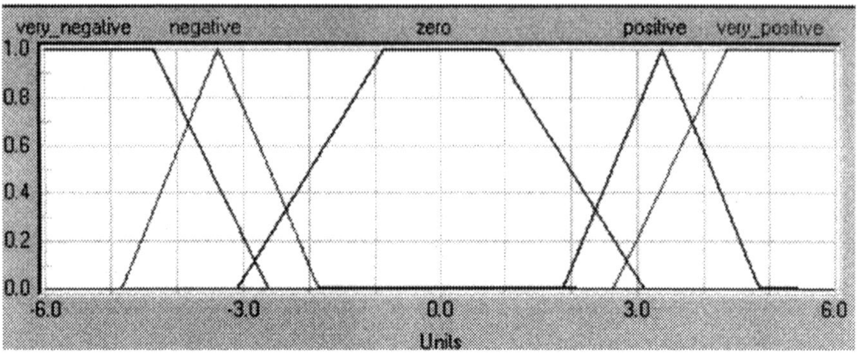

Fig. 5. Membership functions for the error. The unit is meter.

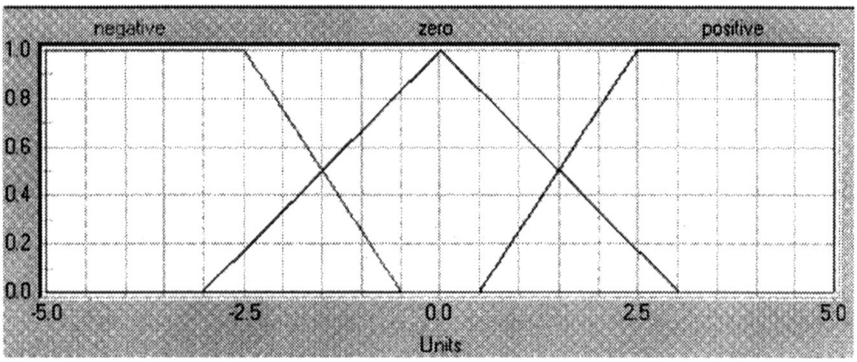

Fig. 6. Membership functions for the change of error. The unit is meter by 4 minutes.

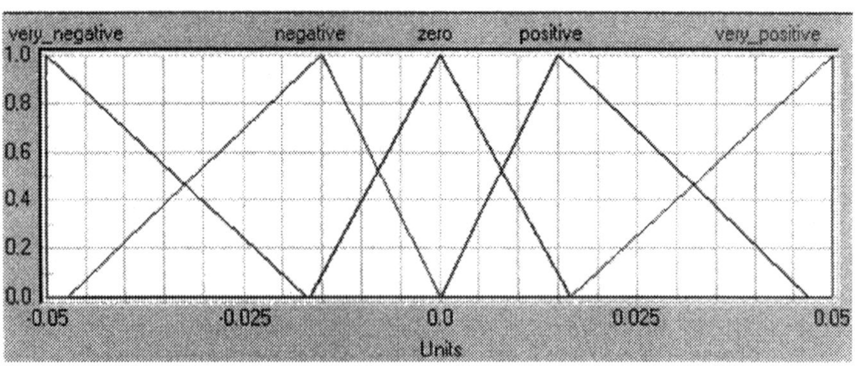

Fig. 7. Membership functions for the speed change. The unit is m/min.

The rule base includes 15 rules. The fuzzy inference is calculated using max-min decomposition. Also all those rules, in which either of two terms is true, are considered with the power 0.1. This kind of method makes the controller more sensitive and stable. Defuzzification is made using the Centre of Maximum method. All membership functions were tuned experimentally.

3.4.2 The Slow Controller

The calculation of input variables for the slow controller is based on the line fitting of the last two hours burn-rising point data using the least squares method. The error is calculated from the linear equation and the change of error from its derivative. The line fitting makes is possible to eliminate periodic disturbances, which are caused by machine wear and material chances. Fitting is used instead of filtering, because it is not sensitive to changing circumstances and disturbances.

The third input to the slow controller is the error calculated by the fast controller. This provides the controller with the newest possible error information and leads to more accurate and tighter control. The layout of the slow controller is shown in Fig. 8. Figs. 9–11 show the membership functions for the input variables and Fig. 12, respectively, for the output.

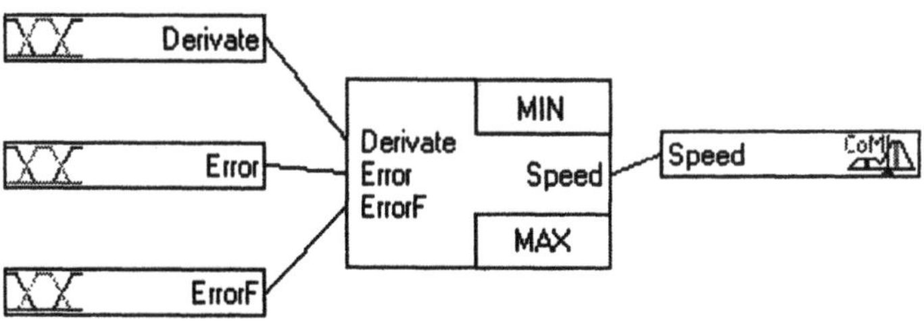

Fig. 8. Layout of the slow controller.

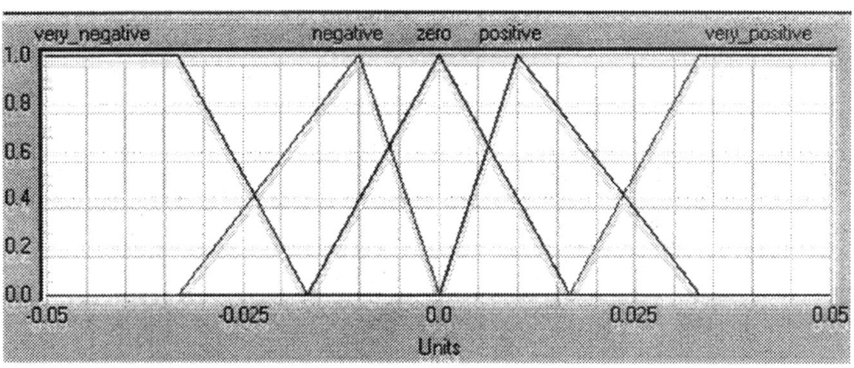

Fig. 9. Membership functions for the change of error. The unit is meter by 2 hours.

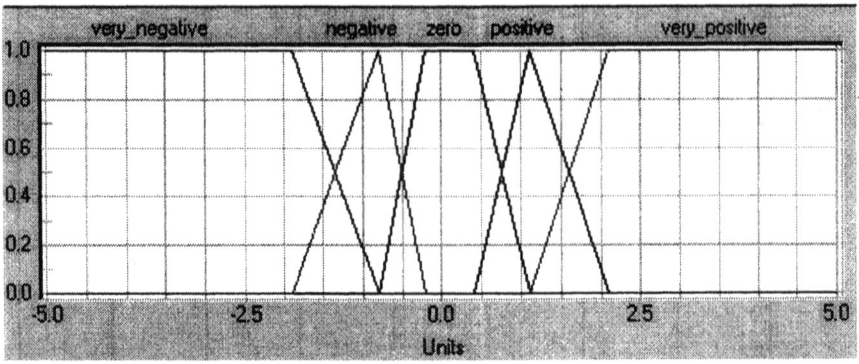

Fig. 10. Membership functions for the error. The unit is meter.

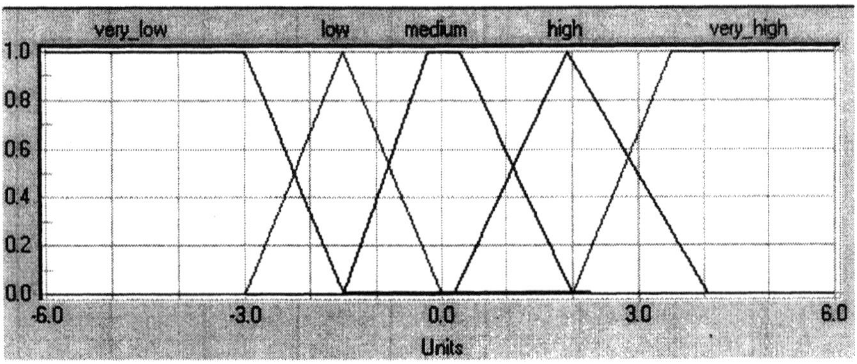

Fig. 11. Membership functions for the fast controller's error. The unit is meter.

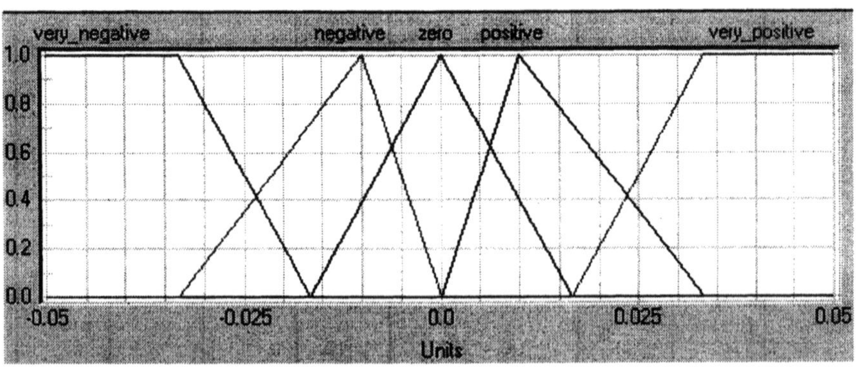

Fig. 12. Membership functions for the speed change. The unit is m/min.

Fig. 10 shows that also in the slow controller the dead band is used around the set point. It is, however, narrower than in the fast controller. There are also more membership functions for each input variable. Therefore the slow controller is able to reach tighter and more accurate control, but with the cost of a bigger rule base. The size of the rule base is not, however, critical from the application point of view.

3.5 Process Test

During the test period, the controller worked correctly. It controlled the strand speed more accurately and more effectively than manual operators during a comparison period. Thus the target of the project was attained. The BRP value and the strand speed value during the test periods of fuzzy control and manual control are presented in Figs. 13 and 14. Usually the fuzzy control executes control actions more frequently than the operators and reacts more quickly to disturbances. As a result, the deviation of the BRP value is smaller and the given limits are not exceeded so often in the case of fuzzy control.

There exists certain fluctuation in the value of the BRP and in the value of the waste gas temperature, respectively. This instability which appears as the periodic fluctuation depends mainly the on inconsistent mix volume of the sintering pallets on the strand and clogging of the gaps between the rods of hearth in some pallets. This causes fluctuation in the bed depth (changes of the distance between hearth and sloping chute) and mix density (changes in the bed compression due to mix volume changes) and thus fluctuation in the burn through time of the bed. This fluctuation is clearly dependent on strand revolutions and complicates the tuning of the controller. The best performance will be attained when the strand speed control will make corrections based on mix quality changes (permeability, coke content) or mix feeding disturbances, which are not periodic ones. There should not be any need for fast periodic changes of the strand speed because it is not the right way to manage sintering by breaking and accelerating the strand speed fast by turns. Homogenising of the quality of sintering pallets and cleaning of the hearth has been carried out and after these actions there have been significantly less fluctuations and the controller could be tuned tighter and faster.

Table 1 shows the results for both controllers compared with the manual control. It shows that the set point is reached with all methods on average, but fuzzy controllers lead to much smaller standard deviation.

Table 1. Results from test periods.

Controller	Calculation method	Average (m)	Standard deviation
Fuzzy test period	Fast controller	25,54	2,10
Fuzzy test period	Slow controller	25,56	1,05
Manual test period	Fast controller	25,55	2,42
Manual test period	Slow controller	25,60	1,12

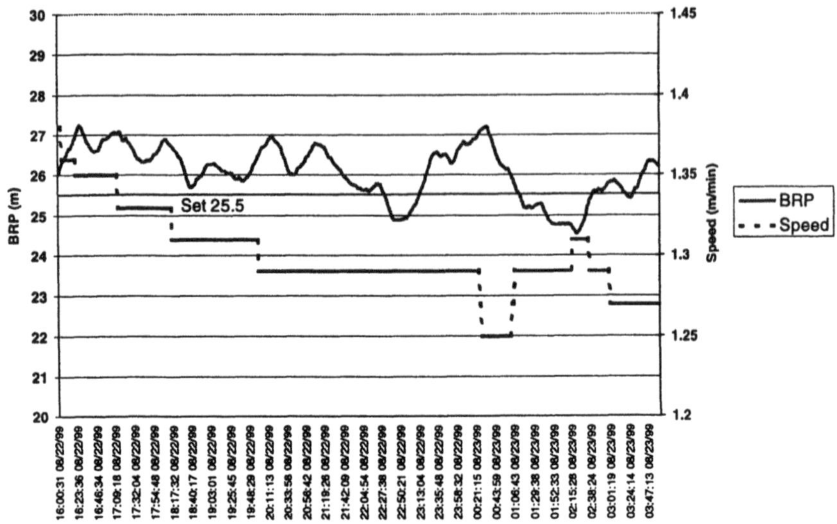

Fig. 13. Fuzzy control test period. The set point for the BRP has been 25.5. Continuous line represents the BRP and the discontinuous one denotes the changes in the strand speed.

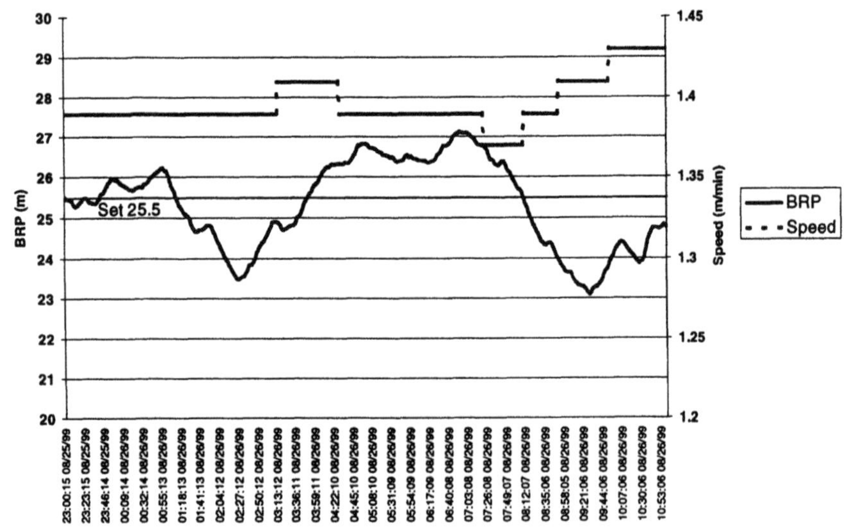

Fig. 14. Manual control test period. Denotations are as before. Note that the strand speed has been all the time on considerably higher level compared with the fuzzy control period. Also a smaller range of the strand speed has been used compared to Fig. 13. Operators have been cautious in their controls.

4 Discussion

The project showed that it is possible to control the strand speed of the sintering machine at least as good as operators at their best, usually even better. It this case, fuzzy logic was a very good choice to make an automatic control system. It was a quite easy to design and develop and it is shown that the developed control system works very well. The problem with sintering process is that the controlled process is so large and complex. Therefore the process itself must be known very well before it is possible to start designing a controller.

The developed controller includes actually three different variables to manage: the position of the burn-rising point, the sinter temperature and the waste gas temperature before the multi-cyclone. Difficulty is that these parameters are handled with the same controlled variable, the strand speed. But problems were figured out during the development and resulting controller structure is very stable and effective.

In the future, tuning of the membership functions will be continued and mathematical handling of the process parameters will be developed. The aim of these changes is to attain more stable control performance and also to take into account the changes in the sinter temperature more perfectly.

5 Summary

The fuzzy strand speed control was developed to manage the burn-through of the sintering bed on a Dwight-Lloyd-type sintering machine. The controller measures the position of the burn-rising point (BRP) along the sintering strand and controls this position by the strand speed. This position is defined in practice as the point in a certain waste gas temperature curve measured from the wind boxes under the strand by recently installed temperature measurement lines.

Sintering process is a complex and multivariable system, which includes long delays. Process is also exposed to different kinds of fault situations and periodic fluctuations. Therefore, a multilevel controller including fast and slow controllers was developed. The fast controller compensates for bigger disturbances and handles unexpected fault situations. The slow controller is a fine tuning controller, which compares the BRP position measured over a long time period with the set point. Also both controllers follow the cooled sinter temperature on a flap top conveyer and the average waste gas temperature under the strand. The set point of the BRP controller is adjusted if one of these limits is exceeded.

During the test period the control worked correctly. It controlled the strand speed more accurately and more effectively than manual operation during the comparison period. Encouraged by good test results, the development work goes on.

The target of the project was to decrease the amount of work needed in process control and also to increase sinter production, decrease fuel rate, prolong plant life, stabilise sinter quality and decrease the amount of return fines and dust.

References

von Altrock, C. (1995): Fuzzy logic and neurofuzzy applications explained. Prentice-Hall Inc., Englewood Cliffs, USA

Arbeithuber, C., Jörgl, H.P., Raml, H. (1995): Fuzzy control of an iron ore sintering plant. Control Engineering Practice 1(12), 1669-1674

Hu, J.Q., Rose, E. (1997): Predictive fuzzy control applied to the sinter strand process. Control Engineering Practice 5(2), 247-252

Jang, M., Cho, S., Lee, S., Kwon, Y. (1998): Belt speed control in sintering plant using neural networks. Steel Research 69(10+11), 398-405

Maki, Y., Takashima, N., Obata, H., Iida, O., Nakashima, K., Sawada, T. (1991): Application of fuzzy theory to ironmaking process. Kawasaki Steel Technical Report 24, pp. 8-10

Myllymäki, P., Poutiainen, J. (2000): Sintering machine burn-through management by fuzzy speed control. In: In: Jämsä-Jounela, S.-L., Vapaavuori, E. (eds.): Future Trends in Automation in Mineral and Metal Processing, IFAC Workshop Preprints. Finnish Society of Automation, Helsinki, pp. 172–176

Nakajima, R., Kurosawa, H., Fukuyo, T., Wada, T., Kimura, R., Ona, M., Yamaoka, Y. (1991): Development of a heat pattern control system for a sintering bed. In: Ironmaking Conference Proceeedings, pp. 363, 366-368

Poutiainen, J. (1999): Burn-through management in a sintering machine by fuzzy speed control. Diploma thesis. University of Oulu, Control Engineering Laboratory (In Finnish)

Pattern Recognition for Blast Furnace Temperature Classification

Henrik Saxén[1], Leif Lassus[2]
[1]Åbo Akademi University, Biskopsgatan 8, FIN-20500, Finland
[2]Oy L M Ericsson Ab, FIN-02420 Jorvas, Finland

1 Introduction

The blast furnace (Fig. 1) is the main process in the world used in producing iron for primary steel production (Biswas 1981; Omori 1987). The furnace is a large refractory-lined counter-current heat exchanger and chemical reactor, where enriched and agglomerated ores, sinter or pellets, are reduced and smelted by hot gases formed by combustion of coke and hydrocarbons. Because of the extreme conditions in the furnace it is very difficult and costly to directly measure internal variables of the process. The internal state must, therefore, be estimated from information at the furnace boundaries, such as measurements of the top gas and hot metal quantities, temperature and chemical composition, blast parameters and wall temperatures.

The wall temperatures are measured by thermocouples fitted in the furnace lining at different vertical levels and on different points on the periphery. Operators usually study these signals carefully, since they may (indirectly) reveal changes in the gas and burden distribution and in the activity of accretions at the wall (Matsuda et al. 1989).

Furthermore, it has been claimed that the shape and the level of the cohesive zone, where the softening of the iron-bearing burden starts, can be deduced from the wall temperatures. However, the large number of other measurements that should be studied simultaneously makes this "manual" approach toilsome and impractical. The interpretation is also complicated by the fact that the wall temperature patterns vary with operation practice and production rate of the furnace. The development of methods for on-line classification of wall temperatures is therefore warranted. Recently, several approaches based on intelligent techniques, have been proposed in the literature (Hirata et al. 1990; Otsuka et al. 1992; Matsuda et al. 1994; Zuo et al. 1998) as shown already in Chapter 2 of this book.

The method described in this paper interprets wall temperature patterns from two industrial blast furnaces, and the promising results inspired us to implement the technique at a Finnish steel works (Lassus and Saxén 1998; Saxén and Lassus 1999; Saxén et al. 2000). The model uses the Kohonen self-organising map (SOM) (Kohonen 1989) to classify the temperature vectors. The feature map also provides a visualisation of the results and may therefore be used as a diagram by which the evolution of the furnace operation point can be tracked.

To make full use of the model, its results have been correlated with key variables of the process, yielding relations that can be expressed as a set of operation diagrams. These diagrams, that further aid the operator in understanding and controlling the blast furnace, will be briefly illustrated at the end of the paper, before some suggestions for future work are given.

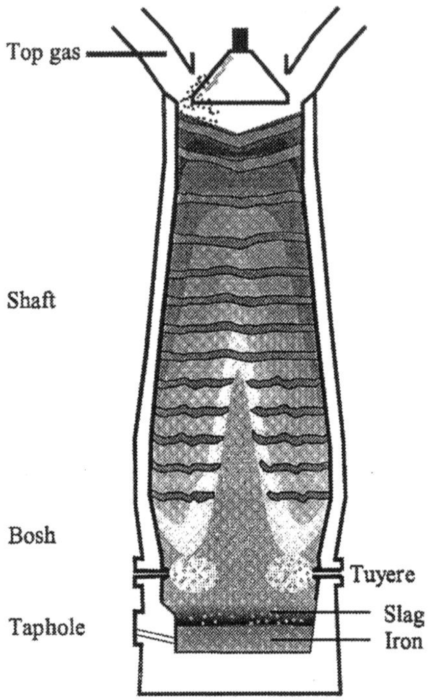

Fig. 1. The blast furnace process.

2 The Model

As mentioned in the introduction, blast furnaces are commonly fitted with thermocouples in the lining, primarily to monitor the extent of the wear of the lining during the furnace campaign. The thermocouple signals also reveal other findings about the conditions at the hot-face of the wall, e.g., changes in the flow conditions of the gas or a possible accumulation of accretions on the wall.

Fig. 2 shows an example of the excursions of the wall temperatures on six different vertical levels during a 20-day period. Clearly, there are changes in the hot-face conditions that can be felt in the measured temperature distribution.

In the blast furnaces studied in this work, wall temperatures were measured on eight vertical levels above the tuyere level (cf. left part of Fig. 3) on different points of the periphery. Since the model to be developed is primarily designed to track slower changes in the operation of the furnace, it was decided to base the classification on one-day averages of the wall temperatures. Averages were also taken on each of the vertical levels; in the applications of the model to be described here, the signals from the uppermost two levels (in the upper shaft) were disregarded. As indicated in Fig. 3, the six-dimensional temperature vectors were classified by the self-organising map.

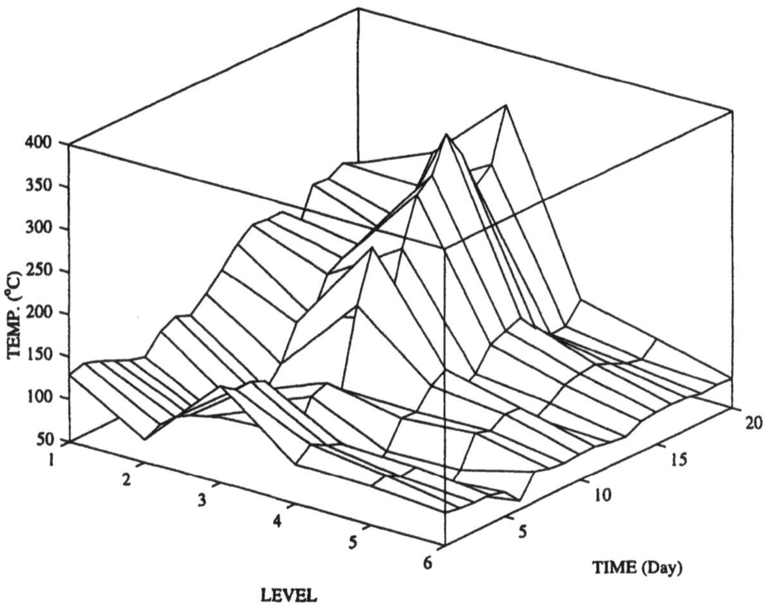

Fig. 2. Example of the evolution of the six wall temperatures. "Level" refers to the temperature measurement points in Fig. 3.

2.1
The Kohonen Classifier

The self-organising map proposed by Kohonen (1989) is a neural network that is trained by unsupervised learning (see also Chapter 1). Every node, *i*, of the network has as many weights (in a corresponding weight vector, $\mathbf{w}_i$) as there are elements in the input vector, $\mathbf{x}$. In this application, the elements of the input vector are the averages of the wall temperatures. Initially, random values are used for the weights, and input vectors are presented, one at a time, but in arbitrary order. For each such presentation, here labelled by a discrete time, t, the node that shows the best match with the input vector, according to

$$i^*(t) = \arg\min_i \{ \| \mathbf{x} - \mathbf{w}_i(t) \| \}, \tag{1}$$

is detected. Next, the weights of this winning node, and its neighbours, are adapted toward the input vector by applying the updating formula

$$w_{ij}(t+1) = w_{ij}(t) + \begin{cases} \eta(t)\left(x_j - w_{ij}(t)\right) & \text{if } i \in \mathsf{N}(t) \\ 0 & \text{otherwise} \end{cases} \tag{2}$$

where η is the learning rate and **N** is a variable defining the neighbourhood (see Fig. 11 in Chapter 1). Both variables should decrease with time during the progress of the training.

Kohonen (1989) states that the neighbourhood should be large initially, e.g., including all nodes, and it should shrink slowly enough to make it possible for the model to arrange itself structurally to the observed patterns. In the end **N** should only include the closest neighbours.

As for the learning rate, it is important that also this variable has big values in the beginning and is reduced as training proceeds, but it is also important that η finally is kept on a low (non-zero) value for a large number of iterations. During this phase, no structural adaptation of the model takes place but the weights are fine-tuned and finally converge.

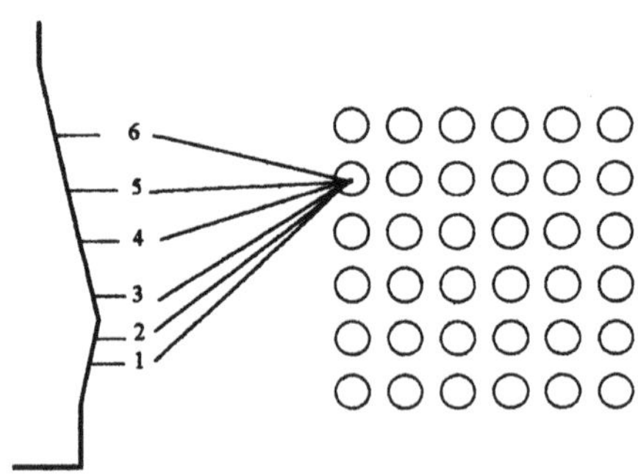

Fig. 3. Schematic illustration of the blast furnace wall temperature classifier. The figure shows also the vertical levels of temperature measurements.

These rules were considered in developing the model to be described. The learning rate and the neighbourhood were given by

$$\eta(t) = \eta_{end} + (\eta_{ini} - \eta_{end})\ e^{-t/\tau} \tag{3}$$

$$\mathrm{N}(t) = \mathrm{N}_{end} + (\mathrm{N}_{ini} - \mathrm{N}_{end})e^{-t/\tau} \tag{4}$$

where t =0,1,2,..., t_{end} is the discrete time and τ is a "time constant". The subscripts "ini" and "end" denote the values at t = 0 and $t = t_{end}$, respectively. Typical values used for the learning rate are η_{ini} = 0.8 and η_{end} = 0.001, while the initial neighbourhood function, N_{ini}, included all network nodes. To achieve the convergence, the total number of iterations should be quite high, and $t_{end} >> \tau$.

The idea of the feature map is to project observations in a high-dimensional space on a lower dimension, still preserving the topological relations between the observations. Therefore, it is expected that similar input (here the wall temperature) patterns would result in winning nodes that are located in the same region of the map. A trial-and-error approach was adopted to select a proper network size; a two-dimensional $N \times M$ node feature map, with $N = M = 5$, was found suitable.

Since the classifier is trained by unsupervised learning, no training set is needed, which is an important feature of a model that is developed for use in a dynamic industrial environment, where model re-tuning is a recurring issue. The drawback of the unsupervised approach is, naturally, that the resulting classification has to be interpreted in order to make full use of the results.

In this application, the feature map may be considered an operation diagram, where the positions of the nodes correspond to specific appearances of the temperature patterns. These, in turn, reflect different internal states of the process. Therefore, conclusions about changes in the inner state of the blast furnace can be drawn by following the evolution of the "activated" regions on the feature map. As the distinct regions of the map may represent different points of operation of the furnace, a tracking of these should be useful for the process operator. However, a problem in studying the evolution of the winning node is that the classification, by definition, "jumps" between discrete states. This may confuse the operators in their interpretation in terms of changes in the state of the furnace. To make the time evolution graded, the measures described below were taken. Denote the row and column of the winning node at time t by $n^*(t)$ and $m^*(t)$, respectively. First, the co-ordinates of the winning node are expressed in the normalised form

$$z_1(t) = \frac{n^*(t) - 0.5}{N}\ ;\ z_2(t) = \frac{m^*(t) - 0.5}{M} \tag{5}$$

on a two-dimensional feature map in the positive unit quadrant. Next, the normalised variables are filtered, e.g. by a first order filter,

$$\overline{z}_l(t) = \alpha \overline{z}_l(t-1) + (1-\alpha) z_l(t)\ ; \qquad l = 1,2. \tag{6}$$

The filtered variables are finally depicted on the feature map, which yields smooth transitions.

2.2 Operation Diagrams

Even though the approach of classifying the temperature patterns is a quite straightforward task, a problem that remains to be solved is that the meaning of the different regions of the map has to be deduced by experience. In order to resolve this dilemma, the (two-dimensional) position of the winning node, ($\bar{z}_1(t)$, $\bar{z}_2(t)$), was correlated with other important process variables, y_i, that characterise the operation of the furnace. To make comparison of the signals meaningful, y_i should be filtered in the same way as the points on the feature map. Using a polynomial of order p to describe the relation we have

$$\bar{y}_i(t) \approx \sum_{j=0}^{p}\sum_{k=0}^{p} a_{ijk}\bar{z}_1^j(t)\bar{z}_2^k(t) \qquad (7)$$

where $\bar{y}_i$ denotes the filtered value of the process variable studied. First, the coefficients, a_{ijk} , are determined by regression. Next, the relation between the "dependent" and "independent" variables can be studied, e.g. by depicting contour lines, i.e. lines corresponding to constant values of $\bar{y}_i$, on the feature map.

It may be noted that the proposed method thus uses a combination of unsupervised and supervised learning techniques. The approach is similar to an extension of the Kohonen SOM with a layer of weights that are trained between the two-dimensional feature map and the dependent variable(s). In linear form, this resembles the action of learning vector quantisation (LVQ) (Kohonen 1989).

3 Illustration Examples

In this section, the model outlined above will be illustrated on data from the blast furnaces of Rautaruukki Steel Mill in Raahe, Finland, for a 2-year period of operation. Different features of the model are illuminated by examples, and, if known, the corresponding changes in the state of the process are reported.

3.1 Long-Term and Short-Term Behaviour

Fig. 4 shows how the method has classified about 200 input patterns for the blast furnace number 1 (BF1). From the figure it can be seen that the method has been able to capture the gradual transitions between the temperature patterns quite well; the profiles that are classified to belong to one group (node) are alike, and there are observations in all classes.

Fig. 5, which shows the corresponding resulting classes for BF2, illustrates that there are major differences between the wall temperature distributions exhibited by the two furnaces. In the case of BF2, there are profiles with high temperatures in the middle of the shaft, while the main heat load in BF1 is in the lower shaft.

This point is thought to correspond to the region above the root of the cohesive zone with a strong flow of gas from the coke slits.

Fig. 6 gives an illustration on how the processed location of the winning node may evolve with time during a time period slightly exceeding one month (using $\alpha = 0.8$ in equation (6)). "S" denotes the starting point and the end point is "E". The temperature profiles are seen to go from quite flat ones through profiles with a higher heat load in the middle of the shaft to finally practically return to the starting point.

The higher temperatures on all levels encountered in the middle of the period reflect a simultaneous increase in the peripheral flow of the gas, which was confirmed by studying the radial distribution of the gas temperature by a probe in the furnace throat. Such a behaviour where a disturbance upsets the operation point for a few weeks, say, followed by a gradual recovery of the original state is typical in the practical operation of the furnace (Karjalahti 1999).

The classifier can be used to optimise blast furnace conditions by following the relation between the key performance indices and the temperature classes on a longer-term basis. In the blast furnace, it is necessary to strictly control the gas and burden distributions in the lumpy (dry) zone to attain an efficient operation with low consumption of reducing agents. The conditions in the peripheral part of the furnace are of central importance, since this region corresponds to a considerable part of the volume of the furnace. This region is also strongly affected by the wall cooling. It should be noted that a too low cooling rate may lead to erosion of the wall, while a too high cooling may result in formation of large accretions that disturb the descent of the burden and reduce the furnace cross-section area and thus the effective volume and productivity.

Table 1 reports a set of key variables of BF2 for four different operation points during a period of approximately one year. The table shows a gradual improvement in the performance, in terms of productivity, reported as tonnes of hot metal (thm) per day, rate of reducing agents (coke and oil) and a sudden dramatic decrease in the specific heat losses.

Fig. 7 illustrates that the improvement corresponds to a clear evolution of the operation point on the feature map. The best state of the furnace in question, using the presently adopted operation practice, is, obviously, to be found in the upper left corner of the map. Therefore, in case of disturbances that perturb the gas distribution and the heat load, counter-measures should be taken to bring the furnace back to a heat-load profile in this region to maintain high productivity and fuel efficiency.

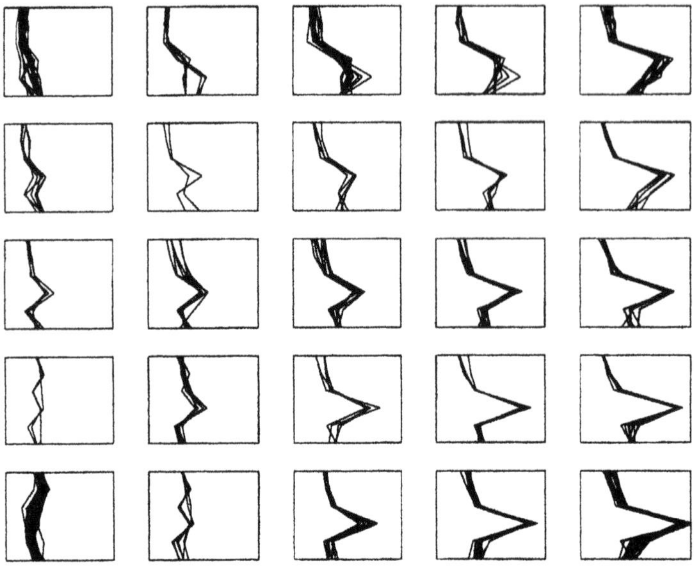

Fig. 4. Example of two hundred classified wall temperature profiles from the blast furnace number 1 at Rautaruukki Raahe Steel mill.

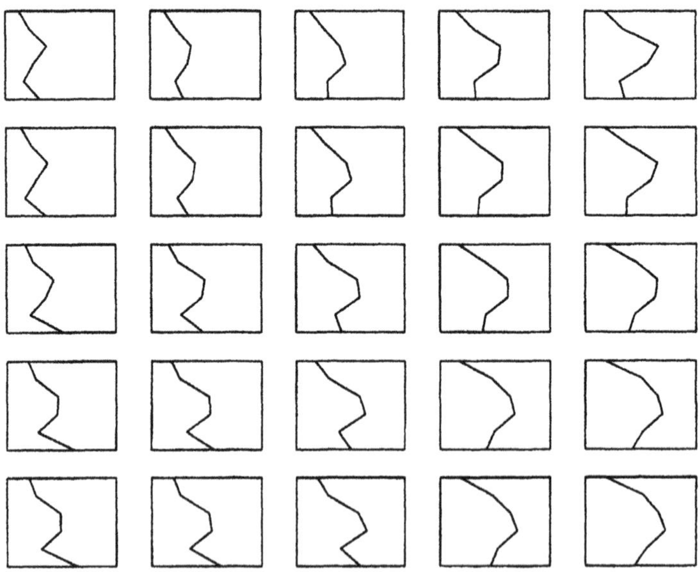

Fig. 5. Classes of wall temperature profiles for BF2 of Rautaruukki Raahe Steel mill.

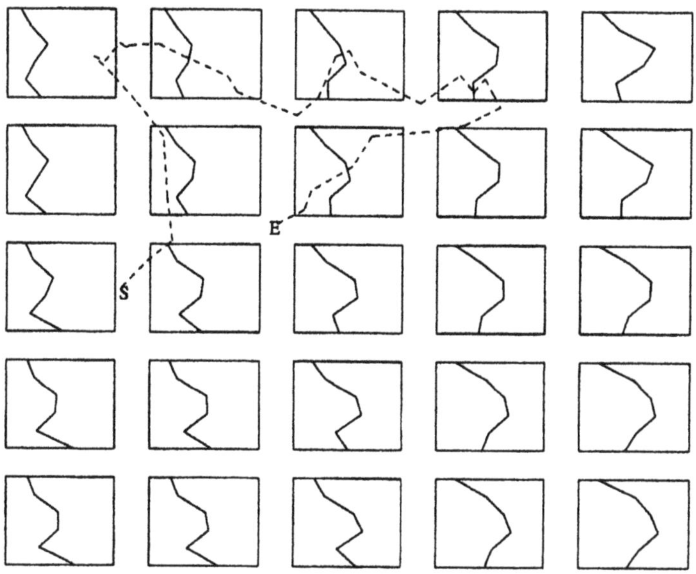

Fig. 6. Filtered evolution of the winning node during a period of about one month in the operation of BF2. "S" denotes the starting point and "E", correspondingly, the end point of excursion.

Table 1. Key variables of performance of BF2 of Rautaruukki Oy, Finland.

Key variable	Feb 1997	Aug 1997	Nov 1997	Mar 1998
Production (thm/d)	2836	2985	3063	3415
Reductant rate (kg/thm)	445	455	450	435
Gas utilization, η_{CO} ( )	0.492	0.491	0.490	0.506
Specific heat loss (MJ/thm)	423	482	486	274
Specific blast volume (m^3n/thm)	1011	1044	1015	937

3.2 Operation Diagrams

In order to make better use of the classification provided by the self-organising map, a set of operation diagrams based on correlation between the wall temperature classes and key process indices was developed by the technique described in sub-section 2. As approximating functions, second-order polynomials (without interaction terms)

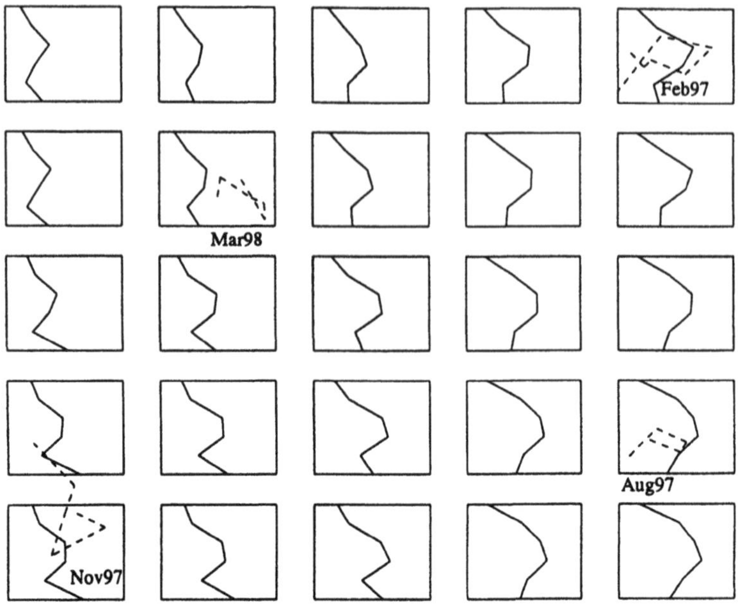

Fig. 7. Points of operation (dashed lines) of the BF2 encountered in February 1997, August 1997, November 1997, and March 1998. Table 1 gives the corresponding key variables of the furnace.

$$\overline{y} \approx a_{00} + a_{10}\overline{z}_1 + a_{01}\overline{z}_2 + a_{20}\overline{z}_1^2 + a_{02}\overline{z}_2^2 \qquad (8)$$

were applied.

The resulting diagram for the hot metal carbon content, which is an important performance and quality index, is shown in Fig. 8.

The figure depicts "iso-concentration" curves for carbon in the interval corresponding to the feature map. The highest values, which are preferable both for hot metal quality and productivity of the steel-processing units down-stream of the blast furnace, are found in the lower part of the diagram.

In Fig. 9, the operation diagram is given for the top gas carbon monoxide utilisation degree,

$$\eta_{CO} = x_{CO_2} / (x_{CO} + x_{CO_2}),$$

where x_i denotes the molar ratio of component i in the top gas. The diagram exhibits a saddle point in the centre and the most advantageous (highest) values are found slightly left of the middle at the top or at the bottom on the figure. These results correspond quite well with the findings of Table 1.

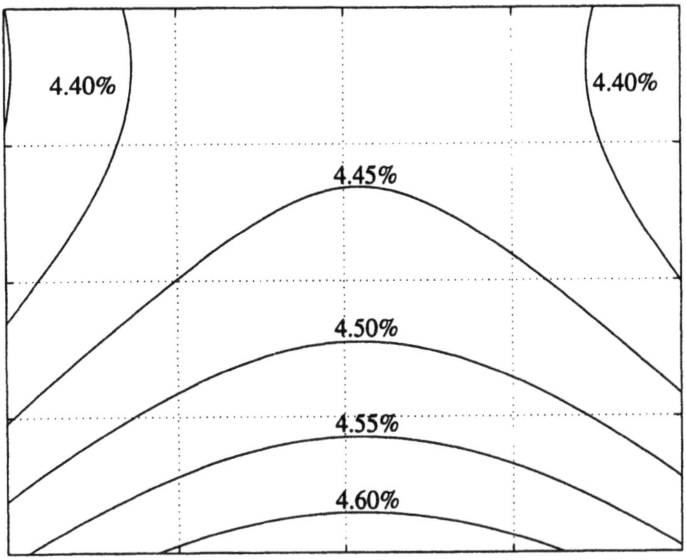

Fig. 8. Operation diagram of hot metal carbon content, C, for BF2.

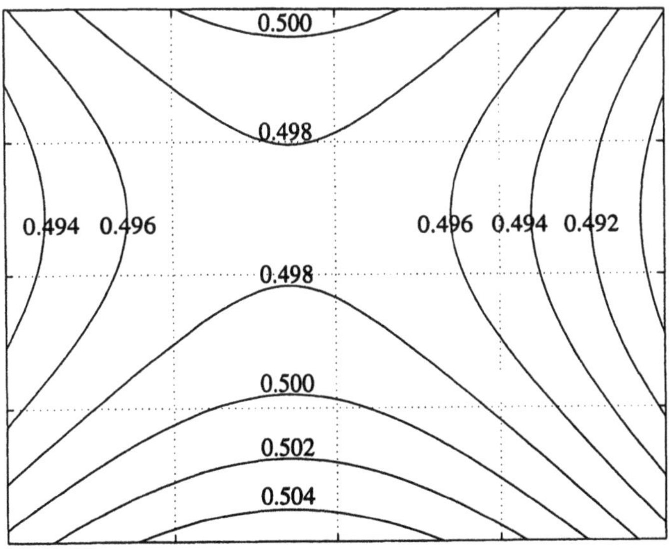

Fig. 9. Operation diagram of utilisation degree of carbon monoxide, η_{CO}, for BF2.

4 Conclusions

This paper has introduced a method for the classification of wall temperatures in the blast furnace, and has illustrated its performance on data from two industrial furnaces. The proposed method is based on a self-organising map, which constitutes a straightforward technique for both the classification and illustration of the temperature patterns on a two-dimensional feature map. In order to facilitate the interpretation of the results, the classifier has been augmented by correlating the classes with key factors that characterise the operation of the furnace.

Since the tool that has been developed makes almost solely use of wall temperature measurements, it is clear that it only supplies partial information about the state of the furnace. Therefore, additional information, extracted by, e.g. mathematical models of gas and burden distribution (Saxén 1996; Nikus and Saxén 1996), should be used in the conjunction with the classifier. By such combinations of different sources of information, it will be possible to provide supporting evidence for the underlying reasons for changes that are observed in the state of the furnace. This would also constitute a logical starting point for further work in the field.

Another possible way to extend the technique is to apply the method to sectorwise temperature patterns (Lassus and Saxén 1996; Saxén et al. 2000) for a more detailed analysis of a possible imbalance in the conditions along the periphery of the furnace.

References

Biswas, A. (1981): Principles of blast furnace ironmaking, Cootha Publishing House, Brisbane, Australia.

Hirata, T., Yamamura, K., Morimoto, S.,Takada, H. (1990): Blast furnace operation system using neural networks and knowledge-base. In: Sixth international iron and steel congress. ISIJ, Nagoya, Japan, pp. 23-27

Karjalahti, T. (1999): Rautaruukki Oy, personal communication

Kohonen, T. (1989): Self-organization and associative memory, 3rd edition. Springer-Verlag, Berlin Heidelberg New York

Lassus, L., Saxén, H. (1998): Tool for classification of blast furnace wall temperature. In: Heidelpriem, H. (ed.): IFAC Symposium Automation in Mining, Mineral and Metal Processing, Cologne, Germany. Pergamon Press, Oxford, pp. 277–281

Matsuda, K., Nagai, N., Konishi, M., Kadoguchi, K., Yabata, T. (1989): Forecasting system for decreasing heat levels in blast furnace. Kobelco Technology Review, 5,1-4.

Matsuda, K., Nakayama, M., Otsuka Y., Konishi, M. (1994): In: Application of AI technology for blast furnace. First International Congress of Science and Technology of Ironmaking. ISIJ, Sendai, Japan, pp. 302-307

Nikus, M., Saxén, H. (1996): On-line model of gas distribution in the blast furnace. Steel Research 67, 121-126

Omori, M. (Ed.) (1987): Blast furnace phenomena and modelling, Elsevier, London

Otsuka, Y., Matsuda, K., Hanaoka, K. (1992): Application of neural network systems to pattern recognition of blast furnace operation data. Kobleco Technology Review, No.15, 12–16

Saxén, H. (1996): Interpretation of probe temperatures in the blast furnace using polynomial approximations. Steel Research, 67, 73-78

Saxén, H., Lassus, L. (1999): Operation diagrams based on blast furnace wall temperatures. 58th Ironmaking Conference, ISS, Chicago, IL, pp. 75–79

Saxén, H., Lassus, L., Seppänen, M., Karjalahti, T. (2000): Pattern recognition and classification of blast furnace wall temperatures. Ironmaking and Steelmaking, 27, 207-211

Zuo, G., Ma, J., Björkman, B. (1998): Some applications of neural networks for prediction of blast furnace irregularities. Steel Research, 69, 41–48

Temperature Control of a Smelting Plant

Tomi Palovaara, Matti Sipola
Rautaruukki Steel, Primary Products, P.O. Box 93, FIN-92101 Raahe, Finland

1 Introduction

The temperature control is an important issue in the steel mill. Tighter quality requirements and the aim of increasing production call for better control of the temperatures in different process stages in a smelting plant.

The main goal of the temperature control is to be able to cast the steel as near the target temperature as possible. Failures in achieving the target cast temperature decrease the quality of the end product and the production capacity. Succeeding in the temperature control in the smelting plant is one of the most important factors when trying to achieve undisturbed production.

There has been a lot of research work done for improving the overall temperature control of liquid steel starting from the blast furnace until the casting.. Various models have been developed to help the operators in their decision making. Especially a lot of attention has been paid to the thermal modelling of steel ladles, which has a significant role in the evolution of steel temperature during ladle treatments. Some selected references are shown at the end of this Chapter.

To be able to achieve the right casting temperature it is essential to obtain the right temperature after the ladle treatments. In Rautaruukki Raahe Steel success in the temperature control is measured by the percentage of casts, which are within ±10°C compared to the target cast temperature. There are also other indicators such as the number of interrupted casts due to failures in the temperature control and the amount of cooling scrap used at the stirring station.

This paper deals with the application, which was made to improve the temperature control in the stirring station and further enhance the control of casting temperatures. Another expectation was that there would be a change in the operating practices so that the operations made in the stirring station would be more stable. This includes the decreasing the use of cooling scrap and maximising the gas stirring times, which have clear economic benefits. It was expected that no equipment investments are needed.

Objectives set for the application fulfilled pretty well. The temperature controls both in the stirring station and casting were improved. Also the usage of cooling scrap decreased. These enhancements were possible because of more stable operation.

2 Process Description

The steel making process at Rautaruukki Raahe Steel (Fig. 1) consists of two mixers, three converters (LD-KG), ladle furnace (LF), vacuum degasser (VD), three stirring stations and five continuous casters.

The pig iron from the blast furnaces is poured into the mixers after desulphurisation and slag skimming. The capacity of the mixers is 1300 tons. Mixers are temporary containers, where the composition and the temperature of the pig iron become even. Usually converters are in operation so that two of the three work at the same time and the third one is under maintenance. The capacities of the converters are 123 tons. The first action in the converters is to choose the heating order, which consists of one or more customer orders. Then the charge model calculates the required amount of hot metal and metal scrap. The calculation is based on the ordered amount of steel, the composition and the temperature of the hot metal, the target temperature and the carbon content of the crude steel at the end of the blow.

First the scrap is charged into the converter. The share of the scrap is about 20 percent of the total charge. Then the hot metal is poured into the converter from the charging ladle. At the beginning of the oxygen blow also lime is added for the slag formation. In the converter, the oxygen is blown from above through a lance and inert gas is fed from below through nozzles. The primary purpose of the oxygen is to oxidise carbon, while the bottom blow improves the mixing.

The oxygen blowing lasts about 17,5 minutes. After that, the bath temperature is measured and the sample is taken. During waiting for the sample result the temperature can be corrected either by adding coolants or by high gas flow through the bottom nozzles if the temperature is too high. If the temperature is too low a re-blow of oxygen can be performed.

The alloy addition model calculates the amount of needed alloys corresponding to the sample result. Alloy additions are made during tapping to a ladle. After tapping the steel ladle is transported to the secondary steel making processes, which are the stirring station, ladle furnace and vacuum degassing.

Main purposes of the gas stirring station are to tune the temperature and composition to the target values and to remove the temperature and composition gradients before casting. The structure and the equipment of the stirring station are shown in Fig. 2.

The stirring treatment is made to every heat, which goes through the station. The stirring occurs by feeding argon gas through porous bricks, which are in the bottom of the ladle. If the bricks are blocked then a lance is used for stirring. The temperature can be tuned by gas stirring or by cooling scrap addition and stirring. The final composition can be tuned via wire feeding, which is used to adjust and complete the alloying made during tapping. The wires have a steel cover and the alloying materials are in the form of a powder.

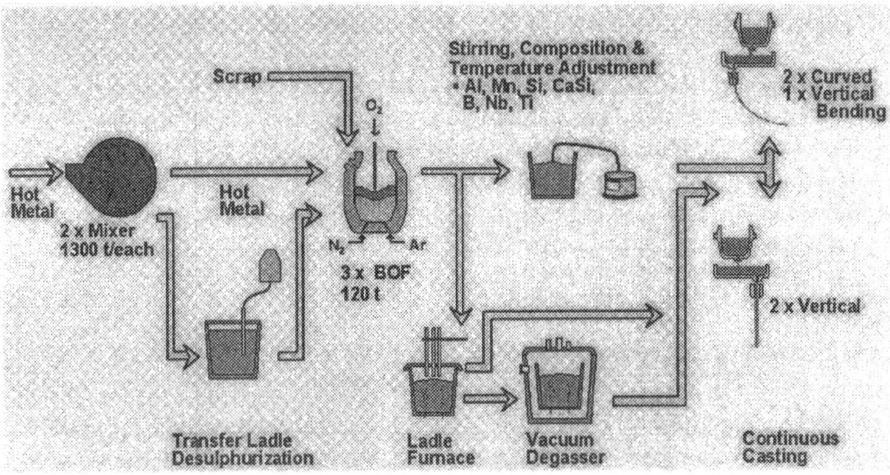

Fig. 1. The steel making process at Rautaruukki Steel.

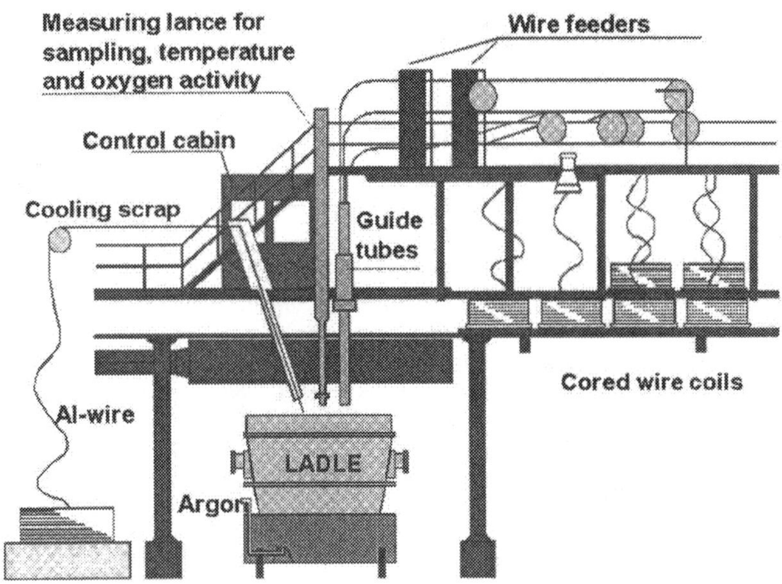

Fig. 2. Stirring station.

Ladle furnace is used for the temperature and composition adjustment and desulphurisation. Heating the steel in the ladle furnace compensates temperature losses due to the vacuum treatment. A vacuum degasser is used for reducing carbon and hydrogen content in steel. After secondary steel making processes the steel ladle is transported by the carriage and crane to continuous casting.

In the casting plant there are five single strand continuous casters. Casting process starts by opening the steel ladle. Then the steel flows in a continuous stream through the ladle shroud into the tundish. From the tundish the steel flows through a submerged entry nozzle into a mould where the solidification of steel starts. Partly solidified cast strand is pulled out from the mould by rollers. The inner part of the cast strand is solidified in the secondary cooling zone by air-water spray cooling. At the end of the casting machine a liquefied petroleum gas oxygen burner cuts the cast strand to slabs.

3 Existing Control Models

In the smelting plant there are two models, which play an important role in the application made for the stirring stations. Those are the temperature control model and the scheduling model, which work under the upper level operating system NEUVO. Next describes how the models work.

The temperature control model produces the target temperatures to every process steps from tapping to casting. Defining of the temperatures occurs in the reverse order compared to the manufacturing route because reaching the right casting temperature is the overall target of the temperature control.

The target temperatures of steel defined by the temperature control model are the liquidus temperature, temperature in the tundish, temperature after the stirring station, temperature after the three minutes stirring and the tapping temperature. The model produces also the target temperatures for the ladle furnace and vacuum degassing.

The temperature control model gets the required information for the calculation from the heat order, alloy addition model, process history data, parameter tables and information given by operators.

The model makes three computing rounds when defining the target temperatures. Choosing the heat order triggers the first round. These target temperatures are updated when the steel ladle for the heat is chosen. The third computing round is done when the alloy addition model has computed the alloy additions.

Next it is shortly described how the target temperatures are computed. Target temperatures for the ladle furnace and vacuum degassing are not dealt with because the application is made for the stirring station. The steel grade determines the liquidus temperature. Target temperature in the tundish is the liquidus temperature plus a certain value from a parameter table, which depends on the steel grade and the type of the casting machine. The target temperature after the stirring station is the target temperature in the tundish plus the predicted cooling of steel from the leaving moment of the stirring station to the temperature measuring moment from the tundish.

The cooling prediction takes the following things into account:

- The mould size, which indicates the casting speed

- The order of the heat in the casting sequence
- Steel ladle parameters.

Following steel ladle parameters are included:

- The time that the ladle has been empty
- Number of casts made by the ladle
- The estimated amount of solidified steel
- Slag in the ladle
- Possible maintenance and heating of the ladle.

Target temperature after the three minutes stirring time is the target temperature after the stirring station plus 14°C. So there should be enough superheat for a ten minutes gas stirring. The target tapping temperature is the target temperature after the three minutes stirring time plus predicted cooling of steel from the start of tapping to the temperature measurement after the three minutes gas stirring. The cooling prediction is determined based on the condition of the tapping hole, the amount of alloy additions, secondary steel making operations and the same steel ladle parameters defined earlier.

The scheduling model produces directive times for different operations in the smelting plant. The operators see the whole scheduling in the form of a Gantt-diagram. Defining the times for different operations executes in the reverse order compared to the actual process.

The scheduling of the heat starts when an operator in the converter control room accepts the heat order to be produced. The heat order contains the secondary steel making operations in the form of different number and letter combinations. It also determines the casting machine by which the heat will be cast. The scheduling model defines the earliest possible cast starting time for the accepted heat order. Then it uses history data of the similar heats and predefined duration of some process steps to make the whole schedule from the mixer to continuous casting.

The most important target time for the stirring station application is the steel ladle opening moment. That is the moment when the previous ladle in the casting sequence becomes empty and the next ladle should be ready for opening. If the next ladle is not on the ladle turret of the casting machine in time, the casting speed or the operating time in the stirring station must be decreased. Reducing the casting speed is more recommendable than shortening the gas stirring time in the stirring station because of the quality of the steel.

The operator at the continuous casting machine estimates the time when the ladle should leave the stirring station. That time is then fed into the NEUVO system. The scheduling model predicts the time when the ladle is supposed to become empty by using the history data of those heats, which were cast by the same machine and the same slab width and thickness. Also the automation system calculates the same ladle emptying moment using information of the casting speed and the amount of steel in the ladle.

If the casting speed changes for example due to superheat of steel then the prediction of the scheduling model would no longer match to the actual ladle emptying moment. The solution is that the scheduling model starts to use the prediction made by the automation system when there is a change in the casting speed. The stirring station application uses this prediction produced by the scheduling model when estimating the time what can be used in the stirring station.

Before defining the problem, which this application tries to solve, the operation at the stirring station is described in detail. In Fig. 3 there is a flow chart of the operations for a typical treatment of a heat. The number for each operation in parentheses is time in minutes from the arriving moment of the steel ladle to the stirring station.

The problem here is the following. After the second temperature measurement at the stirring station, the operator should know what actions there should be done in a certain time period to reach the target temperature after the stirring station as precisely as possible. The optimum way for reaching the target temperature is that only gas stirring is needed. If there is not enough time to cool the steel to the target temperature only by gas stirring then a certain cooling scrap addition needs to be done.

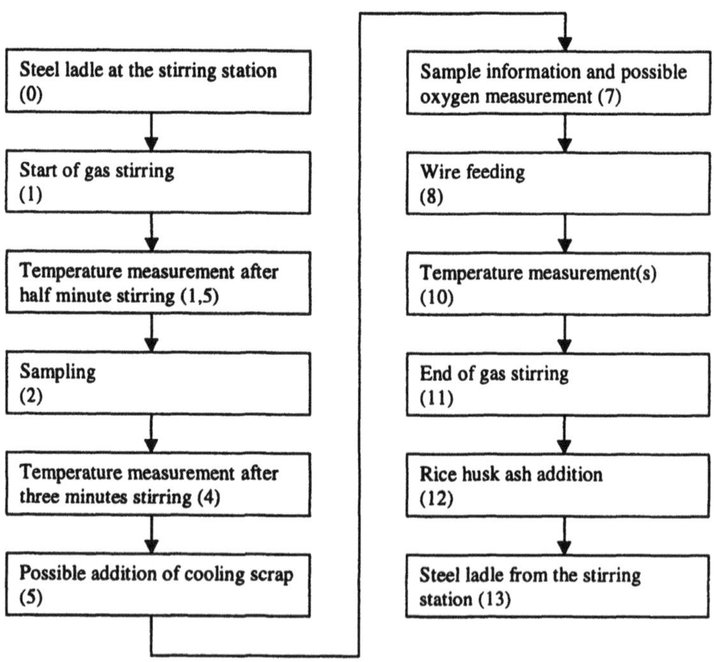

Fig. 3. Flow chart of heat operations. The number in parentheses refers to the time in minutes from the arriving of the steel ladle to the stirring station.

4
Methods Used and the Application

The problem described was solved by a neural network because neural computing is known to be an error tolerant method and a large amount of data was available. A feed forward multi-layer perceptron with a back propagation learning method were used because these were found to be the most used architecture and learning method in the industrial applications.

The main steps used in modelling were statistical analysis of the raw data, pre-processing of raw data, selection of relevant input variables and initialising, training and testing the neural network.

The first thing to do was to find out all the important factors that affect to the temperature control at the stirring station. These factors were found out from the literature and by interviewing operators and the specialists in the area of the temperature control.

Then it was studied what are the most important factors that are already measured from the process so that there will be enough history data for the neural network. The data collected contains 13 actual variables of over 11 000 heats from a half-year period. No more data could be collected because that was all the history databases contained. The reason for such a big amount of data was the knowledge that some restrictions need to be made to ensure the authenticity of the data.

For example the heats, which were re-blown in the converter had to be restricted from the data set because the real tapping temperature is not known for sure. Also the heats that had gone to the ladle furnace or the vacuum degasser and the ones, which contained incorrect or missing values had to be removed. After these and other not mentioned restrictions the number of heats in the data set was about 3700.

In this data set only those heats, which were successfully handled concerning the temperature control could be used in training and testing of the neural network. This means that the temperature after the stirring station may deviate maximum by ±1°C from the target temperature. Pre-processing and adapting of the data set was performed during the whole development period of the neural network.

The tool used in training was Matlab 5.2.0 and its Neural Network Toolbox. The learning methods were back propagation, back propagation with momentum and adaptive learning rate and back propagation with the Levenberg-Marquardt optimisation. Before the training the data set was scaled into the interval of 0.1–0.9 using equation 1.

$$x_n = 0{,}8 \cdot \frac{x_a - x_{min}}{x_{max} - x_{min}} + 0{,}1 \qquad (1)$$

where x_n is the scaled value,
x_a is the original value of the variable,

x_{min} is the minimum value of the variable to be scaled and
x_{max} is the maximum value of the variable to be scaled.

The activation functions used were the sigmoid function in the hidden layers and the linear function in the output layer. In Table 1 there are all the variables, which were used as inputs while developing the neural network.

Table 1. Definition of variables.

1	demand for the cooling
2	target temperature after the gas stirring
3	Temperature after the three minutes stirring
4	Tapping time
5	Tapping temperature
6	The average accuracy of the temperature after the stirring station of the five previous heats
7	The average accuracy of the temperature after the stirring station of the three previous heats
8	The accuracy of the temperature after three minutes stirring
9	Time between end of blowing and start of tapping
10	Empty time of a steel ladle
11	Time between start of tapping and start of gas stirring
12	The accuracy of the tapping temperature
13	Time between start of gas stirring and predicted ladle opening moment

The output of the neural network is the amount of cooling scrap, which is needed to reach the target temperature after the stirring station. 75 % of the data set was used to training and the rest 25 % for testing. After training and testing the output of the neural network was re-scaled using equation 2.

$$x_a = \frac{(x_n - 0{,}1)\cdot(x_{max} - x_{min})}{0{,}8} + x_{min} \qquad (2)$$

Several criteria were used to evaluate the goodness of the neural network:

- The accuracy of the temperature after the stirring station
- Average of the difference
- Correlation between the actual and neural network's amount of cooling scrap
- Minimum, maximum and the standard deviation of the difference between the actual and predicted amount of cooling scrap
- The error term, sum squared error, in training.

To be able to evaluate the accuracy of the temperature after the gas stirring, the difference of the predicted and actual amount of cooling scrap was transformed to correspond with the temperature difference. Transformation was made using the estimation that the effect of one kilogram of cooling scrap per ton of steel is 2°C.

Performance of a neural network depends considerably of the validity of the training data set. That was the reason why the actual amount of cooling scrap could not be used in training if the time in the stirring station was not been used effectively to cooling by gas stirring. According to this, the actual amount of cooling scrap was corrected if necessary. The time between the end of gas stirring and opening the ladle was supposed to be ten minutes.

The next stage of the development process was to define the final input variables of the neural network. All combinations of variables were not tested because some variables were known to be essential for the model. There was a need to pay attention to the recent history. Because of that, a dynamic character was noticed in the form of the average accuracy of the temperature after the stirring station of the three previous heats.

As well as Matlab, BackPack Neural Network System was used in selecting the input variables. By using this tool it was easy to examine the effect of the certain variable to the performance of the whole neural network. The performance was estimated by the root mean squared error. The results concerning the importance of the variables were exactly similar for both tools. The results also corresponded well to the knowledge of the experts.

In total 124 neural networks were trained and tested and the test results were estimated. Different architectures, input variables, learning methods, number of epochs and other parameters were tried in training.

The following parameters were used to obtain the best performance. The training data set contained 1343 heats, which correspond to over one month production through stirring stations. Back propagation with momentum and adaptive learning rate turned out to be the most suitable learning method for this application. Number of epochs in training was 10000. The final architecture was (7,3,7,1), in other words, 7 neurons in the input layer, 3 neurons in the first hidden layer, 7 neurons in the second hidden layer and 1 neuron in the output layer. Fig. 4 shows the final input variables and the structure of the neural network.

Next, the structure and the principle of the model will be described. The computing of the neural network was programmed using C and that forms the base of the stirring station model. The temperature measurement after the three minutes stirring activates the model. The model returns directive values of the amount of cooling scrap and the stirring time. When the model starts, first the parameters and the input variables are called for from the database. Also the target time for the steel ladle opening moment from the scheduling model is fixed. The operational diagram of the stirring station model is in Fig. 5.

The authenticity check of the input variables is made next. Then the neural network computes the amount of cooling scrap. The directive value of the stirring time depends on the demand for the cooling. If the demand is so big that cooling scrap is needed, then all the available time is used for gas stirring. If cooling scrap is not needed, only part of the time is used for stirring. Time, which can be used for stirring depends also on the casting machine and the order of the heat in the casting sequence. Time varies between the moment when the ladle is ready to go at the stirring station and when it is opened in the casting plant. This is a consequence of the different operations of the casting machines and the different line of actions of the operators. This feature was added into the model in the form of IF-THEN rules.

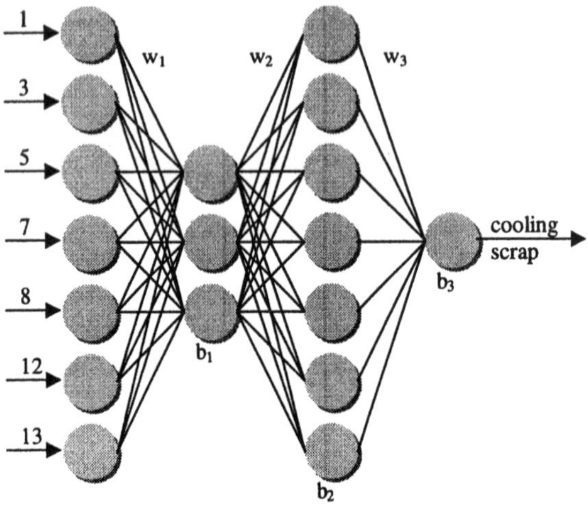

Fig. 4. Structure of the neural network. The numbers refer to the variables shown earlier in Table 1.

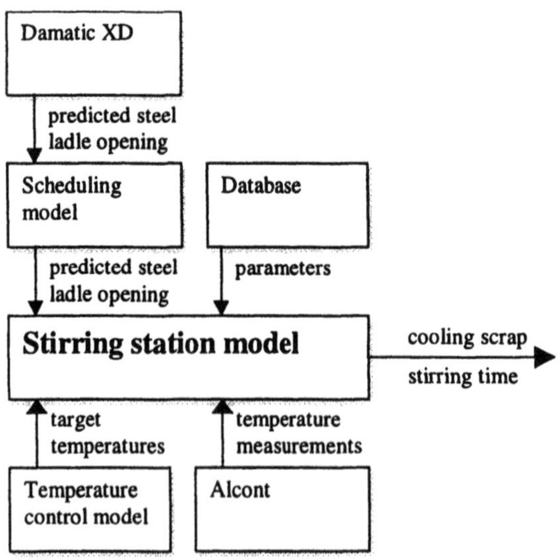

Fig. 5. The operational chart of the stirring station model.

5 Results

Off-line tests were carried out before the implementation of the model. Test data set contained 405 heats, which correspond to ten days production through stirring stations. Fig. 6 shows the result obtained in these tests.

The biggest deviation between the calculated and the actual amount of cooling scrap is less than 180 kilograms. This means a temperature difference of ±3°C. This kind of accuracy in the temperature after the stirring station enables that approximately 90 % of heats would be within ±10°C in casting.

The implementation of the stirring station model was carried out on Feb. 22, 2000. Also the operators were trained how the model works and how to use it.

When estimating the ability of the model to calculate the needed amount of cooling scrap, some relevant factors have to be noticed. The available stirring time has to be utilised and the temperature after the stirring station should be close to the target temperature. The results of the on-line tests in terms of the cooling scrap when the actual stirring time is inside ± 2 minutes compared to the suggestion of the model and the temperature after the stirring station is inside ±3°C compared to the target temperature are shown in Fig. 7. Figure contains 510 heats from the time period of two months.

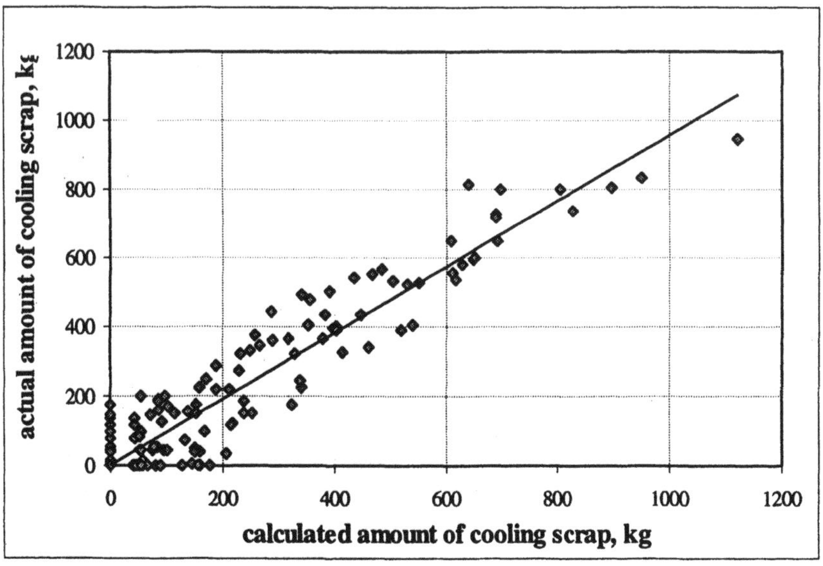

Fig. 6. Results from the off-line testing.

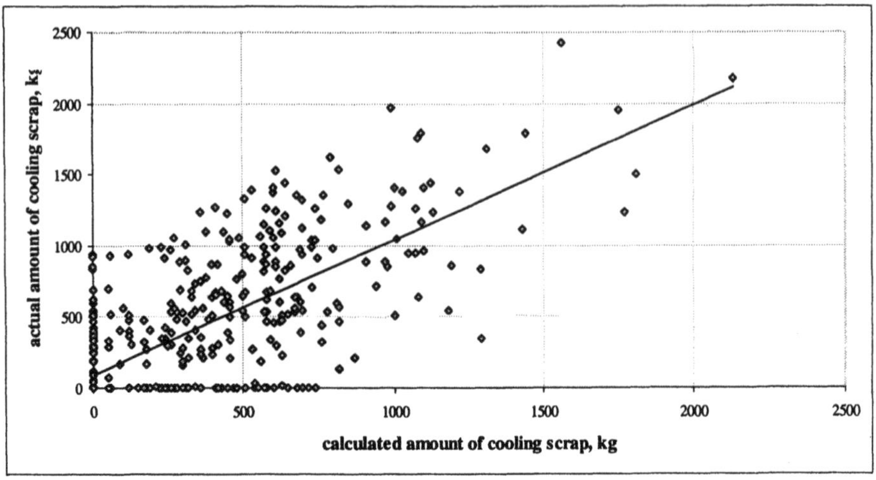

Fig. 7. Results from on-line testing. Modelling of the cooling scrap

The biggest difference between the prediction of the model and the actual amount of cooling scrap is almost 1000 kilograms. The reason why the on-line test results are not as good as the results of off-line tests are discussed in the next chapter.

Fig. 8 shows the average amount of the cooling scrap per heat and the percentage of cooled heats for 6 months period before and 3 months period after the implementation of the model. There are also the averages for the above-mentioned things for the whole period before and after the implementation. It can be seen that the percentage of cooled heats per month has remained quite the same, which means that the operation conditions have been similar before and after the implementation.

When looking at the average amount of cooling scrap used per heat, it is noticed that the consumption has decreased about 50 kilograms per heat after the implementation. The average duration of the gas stirring has remained quite the same it was before the implementation.

There seemed to be some improvements in the temperature control in the stirring station after the implementation (Fig. 9). The improvement can not be big because of the fact that the number of heats, which are already too cold when arriving the stirring station, remains the same from month to month.

The upper curve in Fig. 9 indicates the success in the temperature control in casting. The first two months after the implementation looked pretty good, but May was not so good although the temperatures after the stirring station were behaving very well, which can be seen from the lower curve in Fig. 9. Reason for the weaker development at the casting temperature was one of the five casting machines in which the temperatures were too high during casting. Without this casting machine, 88% of the heats were within ±10°C compared to the target temperature in May. Currently, this kind of problem is solved updating manually the parameters of the temperature control model.

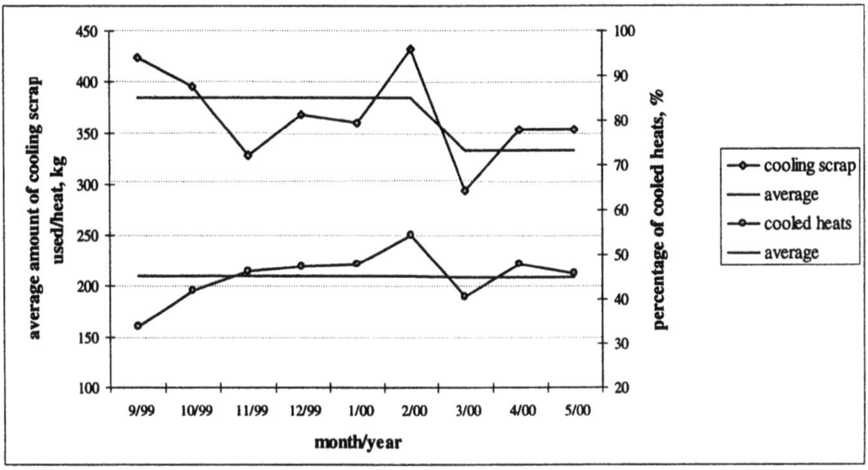

Fig. 8. Results from on-line testing. The development of the cooling scrap and cooled heats. Data is from September 1999 until May 2000. The model was implemented on February 22, 2000.

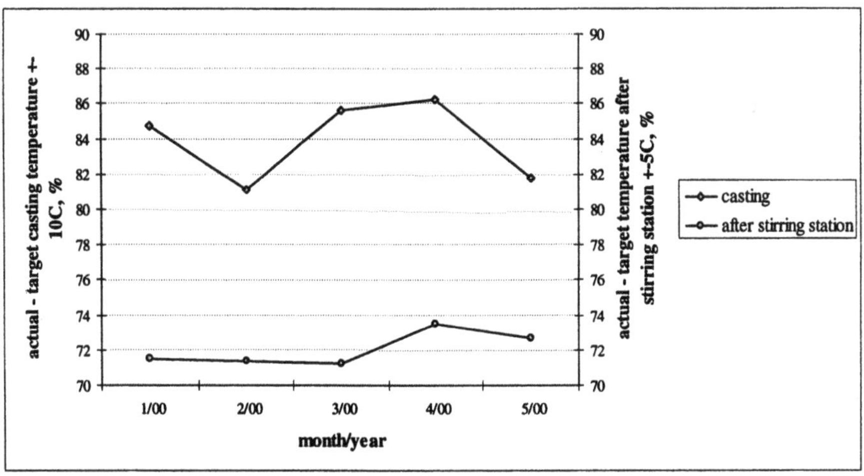

Fig. 9. Results from on-line testing. Temperature deviations from targets in casting and after the stirring station. Data is from January 2000 until May 2000. Once again, the model was implemented on February 22, 2000.

6 Discussion

Objectives set for the application were fulfilled pretty well. The temperature control both in the stirring station and casting were improved. Also the usage of cooling scrap decreased. These enhancements were possible through more stable operations.

There are many reasons why the on-line test results are not as good as the results of the off-line tests. In off-line tests, the cooling effect of gas stirring was assumed to be constant although the intensity of the stirring is adjustable. The biggest individual factor between the off-line tests and on-line use is the difference in scheduling. In off-line tests, the transfer time from the stirring station to the opening moment of the steel ladle was assumed to be constant. In practice the transfer time usually ranges from 8 – 20 minutes and this affects straight to the available stirring time and the amount of cooling scrap needed. This fluctuation is caused by transmission problems from automation system to the scheduling model and the different working habits of the operators.

If the consumption of cooling scrap remains the same as it has been after the implementation of the model, savings would be FIM 0,35 million annually. An improvement in the temperature control makes the quality of the steel better and reduces the number of interrupted casts.

In the future, attention will be paid to make the scheduling more reliable by solving the transmission problems and developing the scheduling model. Also some manual updating of the parameters of the temperature control model is planned to be handled by adding an adaptive sub-model to the temperature control model. This adaptive sub-model is planned to compensate for the changes between ladle treatments and casting and it should improve the temperature control in casting.

7 Conclusions

The new thing that was achieved in this work was the stirring station model, which produces directive values for the operator of the amount of cooling scrap and the stirring time needed. The application was made to improve the temperature control in the stirring station and so, the control of the casting temperature. Another expectation was that there would be a change in the operating practice so that the operations carried out in the stirring station would be more stable. This includes the decreasing the use of cooling scrap and maximising the gas stirring times.

The main steps in modelling were statistical analysis and pre-processing of raw data, selection of relevant input variables and initialising, training and testing the neural network.

The developed model worked well. The process tests showed that the amount of the cooling scrap decreased by 50 kg/heat. Also the temperature control of the stirring station was improved.

References

Austin, P. R., Camplin, J. M., Herbertson, J., Taggart, I. J. (1992): Mathematical modelling of thermal stratification and drainage of steel ladles. ISIJ International 32(2), 196–202

Austin, P. R., O'Rourke, S. L., He, Q. L., Rex, A. J. (1992): Thermal modelling of steel ladles. In: 75th Steelmaking Conference Proceedings. The Iron and Steel Society, Warrendale, PA, pp. 317–323

Barber, B., Normanton, A., Short, M., Patel, A. (1997–1998): Temperature control of liquid steel for casting. Advanced Steel 131–135

Barber, B., Watson, G., Bowden, L. (1994): Optimum ladle design for heat retention during continuous casting. Ironmaking and Steelmaking 21(2) 150–153

Castillejos, A. H., Acosta, F. A., Betancourt, A., Flores, A., Pedroza, M. A., Herrera, M. A., Sepulveda, R. (1997): On-line modeling for temperature control of ladles and steel during continuous thin slab casting. Iron and Steelmaker 24(7), 53–63

Gastón, A., Medina, M. (1997): Thermal modeling of casting ladles: high-alumina, dolomite, magnesite and magnasia-graphite refractories. Iron and Steelmaker 24(1), 29–35

Grip, C.-E. (1994): Measurement of ladle wall temperature to improve control of steel temperature in BOF plant. In: 77th Steelmaking Conference Proceedings. The Iron and Steel Society, Warrendale, PA, pp. 103–115

Herbert, A., Thorndike, M. J., Fawkes, G. (1985): Production of high-quality continuously cast slabs for hotrolled coil and welded-tube application. Ironmaking and Steelmaking 12(1), 30–37

Hlinka, J. W., Miller, T. W. (1970): Temperature loss in liquid steel-refractory systems. Iron and Steel Engineer 8 123–133

Koo, Y. S., Kang, T., Lee, I. R., Shin, Y. K., Gal, H. Y. (1989): Thermal cycle model of ladle for steel temperature control in melt shop and its application. In: 72nd Steelmaking Conference Proceedings. The Iron and Steel Society, Warrendale, PA, 1989, pp. 415–421

Mucciardi, F., Grandillo, A. (1987): Thermal cycling of ladles at Stelco McMaster works. Iron and Steel Engineer 12, 24–30

Olika, B., Björkman, B. (1993): Prediction of steel temperature in ladle through time/temperature simulation. Scandinavian Journal of Metallurgy 22, 213–219

Omotani, M. A., Heaslip, L. J., McLean, A. (1983): Ladle temperature control during continuous casting. Iron and Steelmaker 10(10), 29–35

Perkins, A., Robertson, T., Smith, D. (1986): Improvements to liquid steel temperature control in the ladle and tundish. Fachberichte Hüttenpraxis Metallweiterverarbeitung 24(8), 649–656

Saha, J. K., Ajmani, S. K., Chatterjee, A. (1991): Mathematical model for determining preheating schedule of dolomite ladles in melt shops. Ironmaking and Steelmaking 18(6), 417–422

Short, M. W., Barber, B., Normanton, A. S., Patrick, B., York, R. A. (1997): Temperature prediction and control during continuous casting. In: 56th Ironmaking Conference Proceedings. The Iron & Steel Society, Warrendale, PA, pp. 655–659

Snoeijer, A. B., Boer, R., Hartwig, M., Mostert, R., Böhmer, T. J. M. (1997): Temperature control at BOS no. 2 of Hoogovens Staal. Hoogovens Research & Development 1996–97, pp. 15–20

Szekely, J., Lee, R. G. (1968): The effect of slag thickness on heat loss from ladles holding molten steel. Transactions of the Metallurgical Society of AIME 242, 961–965

Tomazin, C. E., Upton, E. A., Wallis, R. A. (1986): The effect of ladle refractories and practices on steel temperature control. Iron and Steelmaker 13(6), 28–34

Widdowson, R. (1981): Ladle composition and temperature control. Ironmaking and Steelmaking 8(5), 194–200

Zoryk, A., Reid, P. M. (1993): On-line liquid steel temperature control. Iron and Steelmaker 20(6), 21–27

Adaptive Fuzzy Control of a Rotary Dryer

Leena Yliniemi
University of Oulu, Control Engineering Laboratory, P.O.Box 4300, FIN-90014 Oulun yliopisto, Oulu, Finland

1 Introduction

Drying, especially rotary drying is without doubt one of the oldest and most usual unit operations in process industries. Rotary dryers are workhorses, the operation of which is easy and sure, but neither energy efficient nor environmentally friendly. To answer better to the requirements of the modern society concerning working conditions, safety practices and environmental aspects the sophisticated control offers the opportunity to improve dryer operation and efficiency.

The deep understanding of rotary drying is poor, because it is a very complex process including also the movement of solids in addition to thermal drying. So, even today rotary dryers are controlled partly manually based on operators "eye" and experience and partly based on conventional control methods.

The control of a rotary dryer is difficult due to the long time delay of a dryer. Accidental variations in the input variables will disturb the process during long periods of time, before they can be observed in the output variables. For eliminating disturbances in an early stage the interest to more sophisticated control systems as model-based, fuzzy logic, neural nets and their combinations with PID control has increased during the latest years. However, the development of model-based control systems has proved difficult and time consuming due to the complexity of the process. Intelligent control methods based on fuzzy logic and neural nets offer attractive solutions to improve dryer control. These methods make possible to utilise experience, knowledge and historical data, which are much available.

2 Basics of Thermal Drying

When solids dries, two fundamental and simultaneous processes occur

- Heat is transferred to solids from the surrounding medium, air being the most common
- Moisture is transferred as a liquid or vapour within solids and as a vapour from solids surface.

The conventional heat transfer methods are convection, conduction, infrared radiation and dielectric heating, which is one of the modern drying techniques where internal heat is generated by radio or microwave frequencies. In most dryers heat is transferred utilising more than one method. However, each industrial dryer has one predominant heat transfer method and in rotary dryers it is convection where heating is usually provided by direct contact of a hot gas with wet solids.

Due to the complexity of theoretical models of the drying phenomena, the design of industrial dryers is more generally based on the examination of external drying conditions. The drying phenomenon is described by drying curves where the moisture of solids or the rate of drying is presented as a function of time. A typical drying curve for convective drying is shown in Fig. 1.

According to Fig. 1 a drying profile can be divided into three parts: preheating (A – B) (initial drying period), constant rate (B – C) and falling rate period (C – D) or periods depending on the characteristics of solids.

3 Rotary Drying

In process industries large quantities of granular material having particles of 10 mm or larger and not being too fragile or heat-sensitive nor causing other solids handling problems, are dried in rotary dryers. Rotary dryer is one of the most common types of industrial dryers. It is a cylindrical shell constructed usually from steel plates. It is usually slightly inclined, typically 0.3–5 m in diameter, 5–90 m in length and it is rotating at 1–5 rpm with a circumferential speed of approximately 0.2–0.4 m/s. Rotary dryers are usually operated with a negative internal pressure to prevent dust escape. Solids introduced at the upper end will go to the lower or discharge end.

Depending on the arrangement to contact drying gas and solids, a dryer is classified as direct or indirect, concurrent or counter-current. An array of lifting flights of various shapes is constructed inside the shell to shower solids for gas contacting. The necessary heat energy comes from burning gas or oil.

Rotary dryer has two distinct functions: first it operates as a conveyor and second as a heating device. The movement of solids through the dryer is influenced by the different mechanisms shown in Fig. 2: lifting, cascade action, sliding and bouncing.

As the dryer rotates, solids is picked up by the flights, lifted for a certain distance around the drum, and showered through the air as cascading curtain. Most of the drying of solids occurs during this time, solids being in close contact with the gas. The flight action is also partly responsible for the transport of solids through the drum.

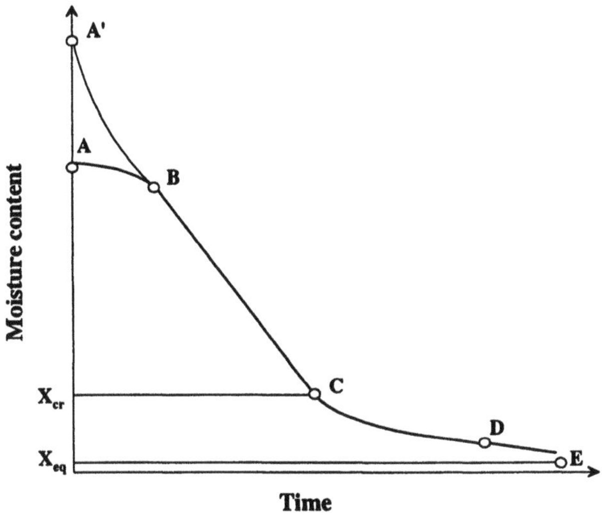

Fig 1. A typical drying curve for convective drying (Strumillo & Kudra 1986).

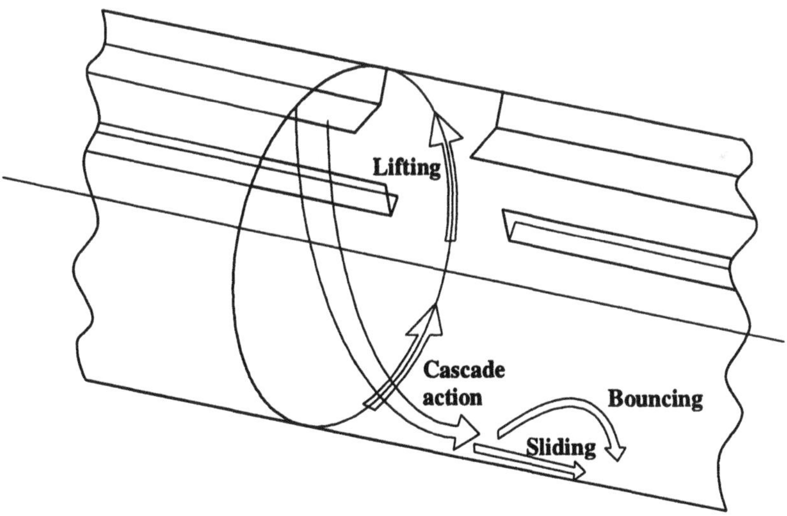

Fig 2. Cascading rotary dryer (Baker 1983).

All the above factors effect on the heat transfer in a drum, and all except the temperatures of solids and drying air have effect on the loading (hold up) and the time of passage of a drum.

4
Control of a Rotary Dryer; a Review

It is difficult to control a rotary dryer due to the long time delays involved. Accidental variations in the input variables as in the moisture content, temperature or flow of the solids will disturb the process for long periods of time, until they are observed in the output variables, especially in the output moisture content. Therefore, pure feedback control is inadequate for keeping the most important variable to be controlled, the output moisture content of the solids, at its target value with acceptable variations. Increasing demands for uniform product quality and for economic and environmental aspects have necessitated improvements in dryer control. Interest has been shown in recent years in intelligent control systems based on expert systems, fuzzy logic or neural nets for eliminating process disturbances at an early stage. These systems add another dimension to the control of a drying process, the human element.

Little development work has been carried out to date on the control of rotary dryers, even though they have been used in industry for many decades. This may be because product quality was not appreciated in the past, because the significance of dryer control for product quality and dryer efficiency was not recognised or because no reliable method for the continuous measurement of moisture content was available. The complexity and highly non-linear dynamics of the drying process also makes it difficult to model a dryer.

The basic objectives for the development of dryer control are

- To maintain the desired product moisture content in spite of disturbances in drying operation,
- To maximise production with optimal energy use and at minimal costs so that the costs of investment in automation are reasonable compared with other equipment costs,
- To avoid overdrying, which increases energy costs and can cause thermal damage to heat-sensitive solids, and
- To stabilise the process.

Manual and automatic feedback systems were the most commonly used methods of dryer control until the advent of process computers in 1970's. In manual feedback control, which is the simplest form, an operator measures the moisture content at the exit from the dryer, mentally compares it with the desired value and adjusts the fuel flow. Automatic feedback control, which is still used in many older plants, is an improvement on manual feedback control in which the moisture content is measured and the fuel flow manipulated. The availability of process computers, better knowledge of dryer modelling and sensor technology have assisted in the development of combined feedforward-feedback control and model-based control in the form of inferential control, internal model control and dynamic matrix control. Interest in developing control systems based on the operator's knowledge and historical data, e.g. expert systems, fuzzy logic and neural nets, has increased in recent years.

The first investigations for dryer control were made in the 1960s and 1970s (Perry 1963; Beck et al. 1971; Harbert 1973,1974; Shinskey 1974; Wright 1976). In most cases the control was based on the control of fuel flow by measuring the output temperature of drying air. Wright (1976) used a continuous infrared moisture analyser to measure the input and output moisture content of the solids and used the results to control the product moisture content via the feed flow or fuel flow.

Pietranski et al. (1982) developed a hierarchical control system for a rotary dryer, with the aim of finding operating conditions that would minimise energy consumption. The system consisted of an expert system supervising a fuzzy logic controller, which acted as the master element in a cascaded control system. Robinson (1989) proposed an inferential control strategy, which is an early model-based approach and is useful if the moisture content cannot be measured directly. Inferential control uses the temperatures of the drying air and the solids together with a process model to determine the moisture content of the solids. Robinson (1992) also developed a temperature-drop control for continuous, convective dryers based on a model in which the product moisture content is related to the drop in the temperature of the drying air during drying and to the dryer speed or feed flow.

Douglas et al. (1992) used two PI controllers, one for the air flow and the other for the rotational speed of the dryer, i. e. controlling the drying time in order to achieve the desired product moisture content. Duchesne et al. (1997) developed and compared five control systems for a con-current rotary dryer, which also incorporated intelligent control methods. The simplest strategy consisted of two feedback loops with PI controllers. The first loop manipulated the fuel flow and kept the output moisture content of the solids at the desired level, and the second manipulated the secondary air flow and kept the output temperature of drying air at the desired level. The second control loop was identical in structure to this but with the addition of a feedforward action based on a phenomenological model, and the remaining three used neural network models: a steady state neural model, a dynamic neural model and a combined PI and neural approach. As the dynamics of solids moisture content are much slower than the dynamics of the drying air, a hybrid controller was developed in which the output moisture of the solids was controlled using a dynamic neural model with one output and the drying air temperature with a PI feedback controller.

Yliniemi (1999) has made the vast research concerning the modelling and control of a rotary dryer. The modelling was based on phenomenological equations. Koskinen (1999) has developed also the model based on fuzzy logic for a rotary dryer. Yliniemi (1999) examined the control of a rotary dryer and developed different control strategies: PI control, model-based cascade control, fuzzy control, neural network control. She also studied two types of hybrid control where fuzzy logic and neural nets have been connected with PI control and compared them with each other by simulations and control experiments with the pilot plant dryer.

5 Adaptive Fuzzy Logic Controllers

5.1 Literature Review

Despite the field of adaptive FLCs is new, a number of methods for designing adaptive FLCs has been developed. In this context the literature review is made, especially three methods which have been applied to the control of the pilot plant rotary dryer are outlined in detail.

Daugherity et al. (1992) described a self-tuning fuzzy controller (STFLC) where the scaling factors of the inputs are changed in the tuning procedure. The process in which the tuning method was applied was a simple gas-fired water heater, since it is widely used in the petrochemical industry and an accurate simulation model is available. The aim was to replace an existing PID controller with a fuzzy controller, using initial guesses as to the fuzzy membership functions and rules to tune the fuzzy controller for optimum performance and to compare the performance of three control regimes i.e. PID, not-tuned FLC and STFLC.

The proposed algorithm was developed and the performance of the three control methods (PID, not-tuned FLC and STFLC) was examined. To assess the robustness of the controller, it was compared with the PID controller when both had 3% zero-mean noise added to the temperature sensor. The tuning was accomplished in 13 steps with the scaling factors initialised to unity. Without some form of adaptive tuning, both controllers came out of critical damping once the process parameters are changed. However, the fuzzy controller was more robust under such changes.

Zheng et al. (1992) presented an adaptive FLC which can emulate not only the control experience of human expert, but also the strategies or thinking of the expert to be utilised in developing a fuzzy controller. The remarkable feature was its dynamic knowledge, which was constructed by a multi-layer rule base. In this case, the lower rule set, called the control rule set, effected the process to be controlled directly as general fuzzy control. The upper layer rule sets, called tuning rule sets, adjusted the control rule set and membership functions of input and output variables. Tuning rule sets were composed of two groups, called the tracking tuning rule set and the specifying tuning rule set, respectively. With tuning rule sets, the control rule set and MFs could be adapted not only for different systems, but also for different control stages of the same system. Upon tuning rule sets, there was a supervisor to manage these tuning rule sets during the real-time operation.

Lui et al. (1994) introduced a novel self-tuning adaptive resolution (STAR) fuzzy control algorithm. One of the unique features was that the fuzzy linguistic concepts change constantly in response to the states of input signals. This was achieved by modifying the corresponding membership functions. This adaptive resolution capability was used to realise a control strategy that attempts to minimise both the rise time and the overshoot. In this approach, the controller rule base did not change, but the definition of the linguistic concepts such as "positive large" adapted constantly to environmental requirements. The thesis was that the resolution (or granularity) of these variables did not need to be constant, but it could change so that fine resolution at the operating point can be maintained, but coarse granularity elsewhere. This approach was applied for a simple two input-one output fuzzy controller.

Jung et al. (1995) presented a self-tuning fuzzy water level controller based on the real-time tuning of the scaling factors for the steam generator of a nuclear power plant. This method used an error ratio as a tuning index and a variable reference tuning index according to system response characteristics for advanced tuning performances. The variability of the reference tuning index was introduced taking into account that the different stage of the system response to an input change has different slit. The method had two different indices for overshoot and undershoot to reduce the number of control intervals for tuning. The values of error and change of manipulated variable were used to determine the grade of instantaneous system performance for scaling factor tuning.

Chiricozzi et al. (1995) proposed a new gain self-tuning method for PI controllers based on the fuzzy inference mechanism. The purpose was to design a FLC to adapt on-line the PI controller parameters.

The essential idea was

- To define a dynamically changing reference trajectory on the error and change of error surface, where they call it "sliding trajectory";
- To compute the error area between the sliding trajectory and the effective one with a special algorithm using this quantity as a performance index to evaluate the system response;
- To draw a fuzzy relationship between the PI gains' correction parameters and the performance index;
- To calculate and to tune PI gains based on the fuzzy inference mechanism.

The aim of the method was to improve the step change response gradually and to assure a certain system overshoot target reached with a reasonable rise time. This algorithm was tested in the PM synchronous motor drive speed control scheme with different set points and extreme initial conditions.

Miyata et al. (1996) proposed a generation of piecewise membership functions in the fuzzy control by the steepest descent method. In this algorithm, MFs of the premise for each rule are tuned independently. Besides only the MFs corresponding to the given input and output data for the learning could be tuned efficiently. Comparing to the conventional triangular and bell-shaped MF's, an expansion of expressiveness was indicated. The learning procedure needed the objective function, inspired to the mean squared error of the controlled output, to be implemented. The tuned values of parameters describing the piecewise linear membership functions (five parameters for each MF) minimised the magnitude of this objective function. Through these adjustments, the desired inference rules could be obtained step by step. This algorithm was tested on a travelling control of a parallel parking of an autonomous mobile robot.

Chen & Lin (1997) presented a methodology to tune the initial membership functions of a fuzzy logic controller for controlled systems. These membership functions of the controller output were adjusted according to the performance index of sliding mode control, so trying to propose a real time simultaneous tuning method. The input variables i.e. output error and change in output error defined this performance index. The general gradient method was adopted to alter the output fuzzy set in the direction of the gradient of the performance index. The idea was to change overall fuzzy controller output under system operation to get small performance index (minimisation problem). In order to reduce the running time of fuzzy controller and simplify the approximate reasoning, a simplified fuzzy approximate reasoning method, which treats the consequent part of control rules as a centre point, was used. In order to assess the tuning fuzzy controller, a non-linear inverted pendulum system and ac oscillating system were selected for numerical examples. The proposed algorithm was not tested on experimental systems, nor any evidence of its robustness is provided.

5.2 Adaptive Fuzzy Tuning for the Pilot Plant Rotary Dryer

Long time delays and settling times, large disturbances in the input moisture of solids, unmeasurable changes in solids, and high non-linearity of the process make the control of a rotary dryer difficult. To eliminate these problems, fuzzy logic has proved its effectiveness. The reason why to develop an adaptive fuzzy logic controller is to design the controller, which adapts to the changes of the process during the operation resulting in robust responses. In the following three different methods for designing adaptive fuzzy logic controllers have been selected for controlling the output moisture of solids in the pilot plant rotary dryer.

5.2.1 Method by Mudi and Pal

Mudi and Pal (1998) presented a simple but robust model for self-tuning fuzzy logic controllers (FLCs). The self-tuning of a FLC was based on adjusting the output scaling factor of a FLC on-line by fuzzy rules according to the current trend of the controlled process. The rule-base for tuning the output SF was defined based on the error and the change of error of the controlled variable using the most common and unbiased membership function. The error was taken as the difference between the set point and the output controlled variable. The proposed self-tuning technique was applied to both PI- and PD-type FLC by making the simulation analysis for a wide range of different linear and non-linear second order processes including a marginally stable system. The performance of the proposed STFLC was compared with the corresponding conventional FLC in terms of several performance measures such as overshoot, settling time, rise time, integral absolute error (IAE) and integral-of-time absolute error (ITAE) in addition to the responses due to stepwise set point changes and load disturbances.

In determining the rules it is important to take into account the following:

- Controllers produce the lower overshoot and reduce the settling time. The controller gain is set small when the error is big but e and Δe are of opposite signs i.e. IF e is PB and Δe is NS THEN α is VS. To minimise the effects of the delayed control action due to the inherent process dead time or measuring lag, the small gain is essential to maintain the performance within the acceptable limit, especially when the process dead time becomes considerably big. When the error is big but e and Δe are of the same sign, the gain should be made very big to prevent the additional worsening situation (i.e. IF e is PB and Δe is PS THEN α is VB).
- Depending on the process trend, there should be a wide variation of gain values around the set point i.e. when the error is small to avoid large overshoot and undershoot i.e. IF e is ZE and Δe is NM THEN α is B, in order to prevent the upward motion from the set point more severely resulting in a smaller overshoot;
- To eliminate any load disturbances the gain of the controller should be as high as possible especially around the steady state conditions (i.e. IF e is NS and Δe is NM THEN α is B). However, at the steady state (e and $\Delta e \cong 0$) the controller gain should be very small (i.e. IF e is ZE and Δe is ZE THEN α is ZE).

For tuning the proposed controller the following steps are used:

- Tune the SFs of the STFLC without tuning the gain and assume that α=1 for a given process to achieve a reasonably good control performance. Because the well-defined method for the determination of SFs does not exist, the suitable values of the SF's are selected from the knowledge of the process and sometimes through trial and error. Finally the good controller without self-tuning is as result. The output of the STFLC is nearly three times greater than that obtained in step 1 keeping the values of the input SFs same as the values of the conventional FLC. This is found experimentally as the aim to maintain the same rise time as the conventional FLC.
- Fine tuning the rules for α depending on the type of the response wanted to achieve. For example if it is wanted further to reduce the overshoot at the cost of increased rise time then the value of α should be kept very small up to medium values of e (i.e. IF e is PM and Δe is NS THEN α is VS, and not simply S).

Design procedure for the pilot plant dryer

Following the method by Mudi and Pal the supervisory controller tunes the output scaling factors of the low-level controller using a gain updating factor. Furthermore the inputs of the supervisory controller are selected the same as the ones of the low-level controller. This reduces the amount of memory required for the database. Next procedure assumes the hybrid PI-FLC (Yliniemi 1999) as the low-level controller. Next chapter shows also results with direct FLC as a low-level controller. In this case the definition of linguistic variables and the rule base is different (Cammarata and Yliniemi 1999).

The design procedure of the STFLC to the pilot plant rotary dryer is as follows:

Step 1: System functional requirements. As the previous studies by Yliniemi (1998, 1999) have proved the FLC applies well to the control of the rotary dryer. The aim of the STFLC is to improve the performance of the existing fuzzy controllers by utilising the operator's expert knowledge.

Step 2: Definition of system parameters. The main controlled variable is the output moisture of solids and the main manipulated variable is the input temperature of drying air, which correlates with the fuel flow. The velocity of solids, which correlates with the rotational speed of the screw conveyor can be used as an auxiliary manipulated variable. The main disturbances are the variations in the input moisture of solids and the feed flow.

Step 3: Definition of the fuzzy tuner. The input variables for the STFLC representing the contents of rule antecedents are as follows:

- Error in the output moisture of solids, e (***input 1***) with the following linguistic labels; e = {VN, N, M, P, VP}
- Change in the error of the output moisture of solids, Δe (***input 2***) with the following linguistic labels; Δe = {N ZE P}

The controller output representing the contents of the rule consequent is:

- Gain updating factor, α (***output***) with the following linguistic levels; {ZE, S, M, B, VB}

The membership functions for all variables have been determined experimentally and are shown in Figs. 3–5.

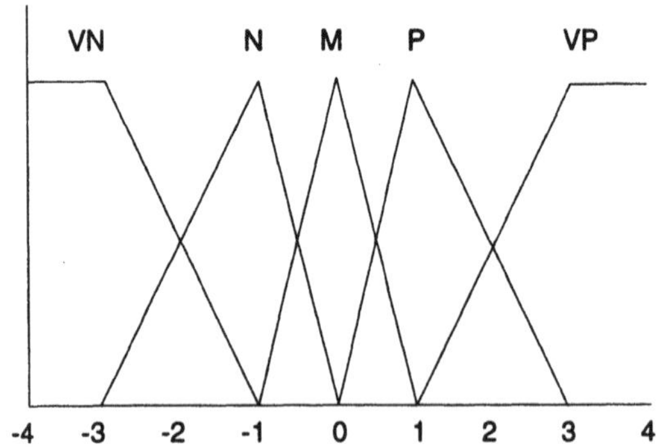

Fig. 3. Membership functions for the error in the output moisture content.

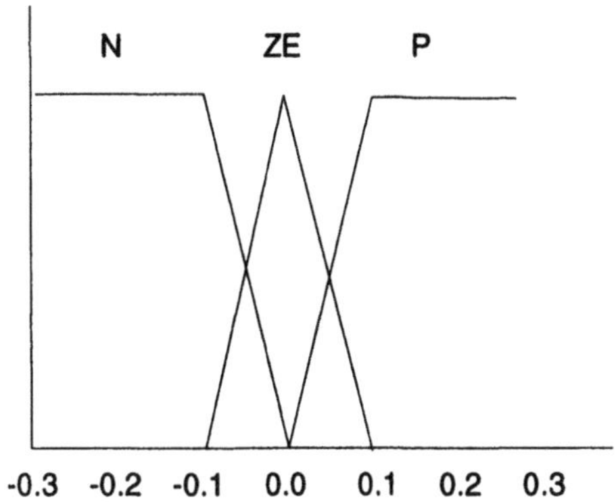

Fig. 4. Membership functions for the change of error.

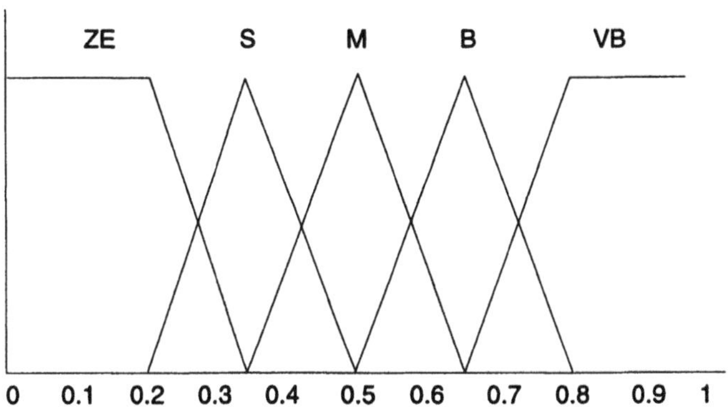

Fig. 5. Membership functions for the gain updating factor.

Step 4: Formulation of the gain updating rules. The rules based on the analysis of the controlled response and on the experience are shown in Table 1.

Table 1. The rule base for gain updating.

E\Δe	N	ZE	P
VN	VB	M	S
N	VB	B	S
M	B	ZE	B
P	S	M	B
VP	S	M	VB

In this application the fuzzification is based on the min-max algorithm, and the inference engine or rule firing is based in the one-level forward data-driven inference, which is the generalisation of the modus ponens rule of inference.

Step 5: Selection a method of defuzzification. The centre of area method (COA) is used to transform the output of the fired rules into the crisp value.

Step 6: Simulation tests. The controller has been implemented in Matlab®, using Simulink and Fuzzy Logic Toolbox. Simulations have been carried out using the hybrid PI-FLC and the direct FLC as low-level controllers developed by Yliniemi (1999) and making step changes in the main disturbance variable i.e. the input moisture of solids.

5.2.2 Method by Chung et.al.

Chung et al. (1998) propose a self-tuning PI-type fuzzy controller with a smart and easy structure. The tuning scheme allows tune the scaling factors by only seven rules. It is known that the PI-type fuzzy controller has good performance at steady state. However, its' shortcoming is long rise time and long settling time. The aim of the adaptive FLC is to overcome these shortcomings.

The structure of this controller consists of two fuzzy logic controllers (Fig. 6): one is a PI-type fuzzy controller at low level directly applied to the process; the other is the fuzzy supervisory tuner controller which adjusts the scaling factors of each MF of the low level controller. This means that the self-tuning controller adjusts three scaling factors for the three linguistic variables of the PI-type fuzzy controller, i.e. G_e (scaling factor of error) G_{ce} (scaling factor of change of error) and G_{cu} (scaling factor of change of manipulated variable).

The requirements are smaller overshoot, faster rise time, smaller steady- state error and smaller load effect. Since the PI-type fuzzy controller is used the steady state performance is certainly better. The strategy is that the system has a positive large acceleration at the beginning. In other words the scaling factor is increased such that the rise time and settling time are reduced. On the other hand the system has a negative acceleration when the output signal is near the set point, i.e. the scaling factors have to be reduced such that the overshoot is reduced or prevented. The seven tuning rules constructed use ratio error (i.e. current /(set point - initial system output)) as input variable and variation of scaling factors of e, ce and cu as output consequent variables. For each linguistic variable seven labels are defined. MFs for the output variables of the FLC tuner controller are the same. The actual scaling factors are obtained using formula, which correlates to the incremental factors. In this formula w is the convergent coefficient and if it is wanted to get faster rise time it would be increased.

Design procedure for the pilot plant rotary dryer

Considering a discrete-time fuzzy controller with two inputs and a single output, the error and change in error of the process as its inputs are defined as:

$$e(k) = sp(k) - y(k) \tag{1}$$

$$ce(k) = e(k) - e(k-1) \tag{2}$$

$$ce(k) = y(k-1) - y(k) \text{ when } sp(k) = sp(k-1) \tag{3}$$

where sp and y denote the applied set point input and plant output respectively. Indices k and $k-1$ indicate the present state and the previous state of the system respectively.

The structure of the self tuning fuzzy controller is shown in Figure 6 where:

$$\begin{aligned} se(k) &= [sp(k) - y(k)] \times Ge(k) = e(k) \times Ge(k) \\ sce(k) &= [e(k) - e(k-1)] \times Gce(k) = ce(k) \times Gce(k) \\ cu(k) &= scu(k) \times Gcu(k) \end{aligned} \tag{4}$$

and $Ge(k)$, $Gce(k)$ and $Gcu(k)$ are the scaling factors.

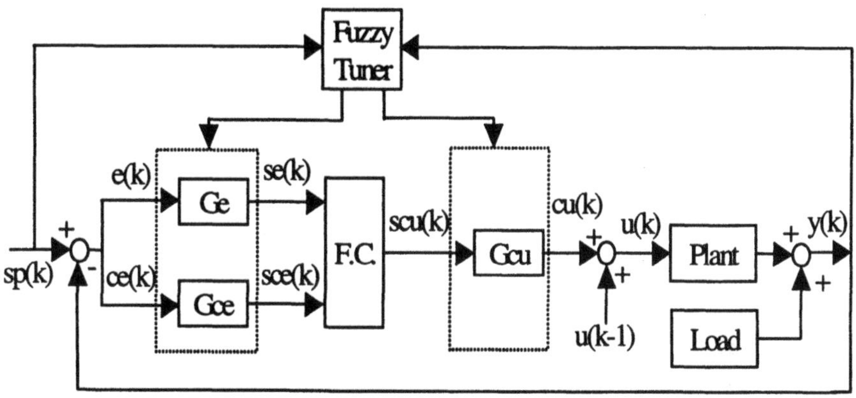

Fig 6. The self-tuning fuzzy controller.

The procedure for the adaptive FLC goes as follows:

Step 1: Definition of system parameters is the same as before.

Step 2: Definition of the fuzzy controller. The input variables representing the contents of the rule antecedents are denoted as follows:

- Error in the output moisture content of the solids, se(k) (***input 1***) with the following linguistic labels: se(k) = {NB, NS, ZE, PS, PB}.
- Change in error in the output moisture content of the solids, sce(k) (***input2***) with the same labels as input 1.
- The control output representing the contents of the rule consequent is the incremental change in the input temperature of drying air, scu(k) (***output***) with the following linguistic labels: scu(k) = {NB, NM, NS, ZE, PS, PM, PB}

Triangular membership functions are shown in Figs. 7 and 8. These are the same as in the original article by Chung et al. (1998).

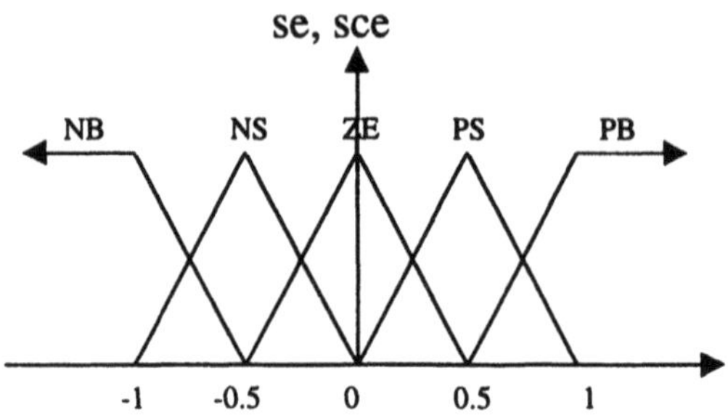

Fig 7. Membership functions for the input variables.

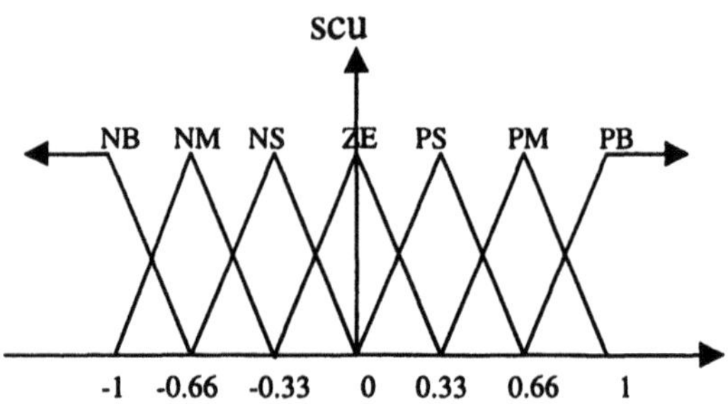

Fig. 8. Membership functions for the output variable.

Step 3: Formation of the control rules. The rule base of the fuzzy controller is shown in Table 2.

Table 2. The control rule base for the adaptive fuzzy controller.

se\sce	NB	NS	ZE	PS	PB
NB	NB	NB	NM	NS	ZE
NS	NB	NM	NS	ZE	PS
ZE	NM	NS	ZE	PS	PM
PS	NS	ZE	PS	PM	PB
PB	ZE	PS	PM	PB	PB

With max-min inference method and the weighted average defuzzification procedure the incremental control signal is obtained. From this the control signal is calculated as follows:

$$u(k) = u(k-1) + cu(k) \tag{5}$$

Step 4: Definition of the tuning variables.

Tuning input variable R is defined as $r = |e/(sp - y_0)|$; y_0 is initial condition, *e* is error, *sp* is the set point and r has the following linguistic labels: r = {VS, S, M-, M+, B, VB}

Tuning output variables are incremental values of scaling factors, *dGe*, *dGce* and *dGcu*. They have seven linguistic labels as {NB, NM, NS, ZE, PS, PM, PB}. Figs. 9 and 10 show the corresponding membership functions.

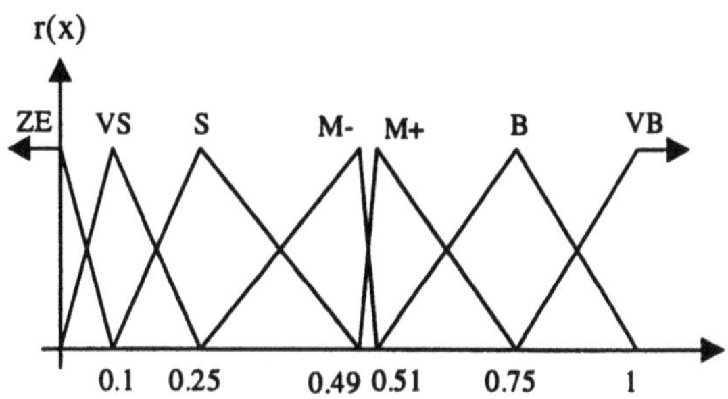

Fig. 9. Membership functions for the input tuning variable.

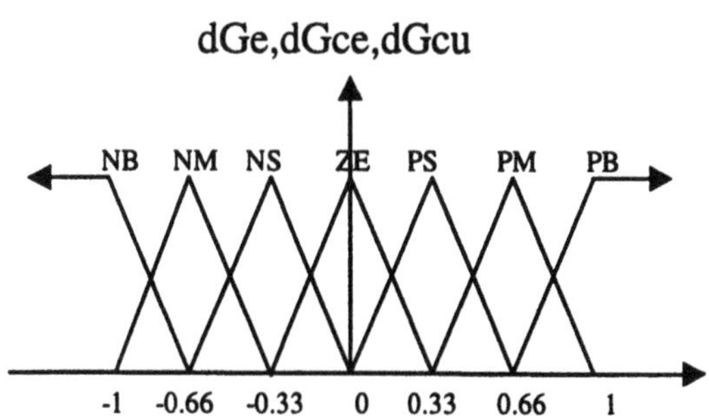

Fig. 10. Membership functions for the output tuning variables.

Step 5: Definition of the tuning rule base. The tuning rule base consists of seven rules as shown in Table 3 (Chung et al. 1998).

Table 3. The tuning rule base.

R	ZE	VS	S	M-	M+	B	VB
dGe	ZE	ZE	NS	NM	ZE	PM	PB
dGce	ZE	ZE	NS	NM	ZE	PM	PB
dGcu	ZE	ZE	NS	NM	ZE	PM	PB

The new scaling factor for the present state can then be calculated as from the incremental factors as follows:

$$
\begin{aligned}
Ge(k) &= \frac{1+dGe(k)}{h} = \frac{1+T(r(k))}{h} \\
Gce(k) &= \frac{1+dGce(k)}{h}\times 10 = \frac{1+T(r(k))}{h}\times 10 \\
Gcu(k) &= (1+1.5\times dGcu(k))\times h\times w = (1+1.5\times T(r(k)))\times h\times w
\end{aligned}
\tag{6}
$$

where $h = |sp - y_0|$, and w is the convergent coefficient. The effect of the parameter h is to normalise each input over the interval [-1 1]. From experience it is known that *Gce* is ten times as large as *Ge* and *dGcu* 1.5 times to get *Gcu*.

Step 6: Selection of a method of defuzzification. The centre of gravity/area method is used to transform the outputs of the fired rules into crisp outputs.

In order to achieve the best possible control there are a number of constants, which may be altered in this method. The one chosen to vary in this application is the convergent coefficient w. The results of the outlet moisture content of the solids for step changes from 2.5% to 3.5% for three different values of w are shown below in Figure 11. For simulation studies the value of w=3 was chosen as it gives the least oscillation.

5.2.3 Method by Ramkumar and Chidambaram

The adaptive fuzzy method developed by Ramkumar and Chidambaram (1995) has the basic idea to parameterise Ziegler-Nichols like tuning formula by two parameters α and β and then to use an on-line fuzzy inference mechanism to tune the parameters of a PI controller. This method includes the fuzzy tuner with the error in the process output as input and the tuning parameters α and β as outputs.

The idea to use the Ziegler-Nichols formula is originally presented by He et al. (1993) who presented the scheme with a single parameter α. In their method, the fuzzy tuning mechanism with the process output error and error rate as inputs adjusts α in such a way that it speeds up the convergence of the process output to a set point and slows down the divergence trend of the output from set point.

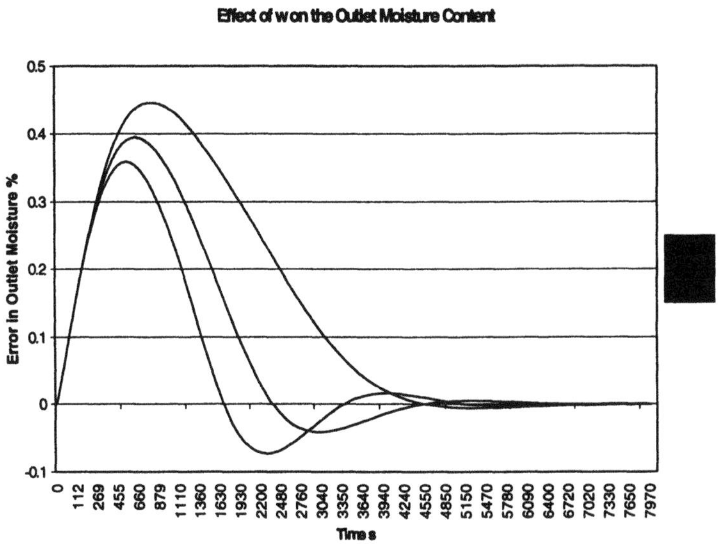

Fig 11. Effect of w on the output moisture of solids. In the uppermost curve w=3, in the middle w=5 and in the lowermost curve w=7.

Design procedure for the pilot plant rotary dryer

The design procedure of the adaptive FLC includes the following determinations:

Step 1:Definition of system parameters is the same as before.

Step 2: Definition of system parameters in terms of fuzzy sets. The input variable representing the contents of the rule antecedents is:

- Error in the output moisture content of the solids, e (***input***) with seven linguistic labels: e = {NB, NM, NS, ZE, PS, PM, PB}

The control outputs representing the contents of the rule consequent is:

- Tuning parameter α (***output 1***) with five labels: α = {VS, S, M, B, VB}
- Tuning parameter β (***output 2***) with three labels: β = {S, M, B}.

The membership functions were determined experimentally (Figs. 12–14).

Step 3: Formation of the tuning rules. Also in this case the tuning rule base consists of only seven rules as shown in Table 4.

Table 4. The tuning rule base.

E	NB	NM	NS	ZE	PS	PM	PB
α	VB	B	M	VS	M	B	VB
β	B	M	S	S	S	M	B

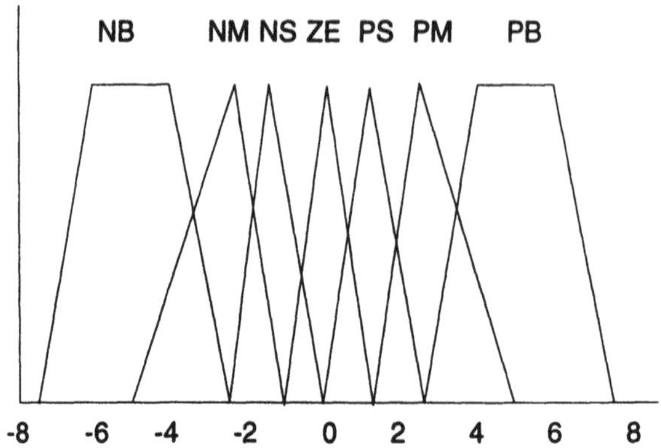

Fig. 12. Membership functions for the tuning input variable.

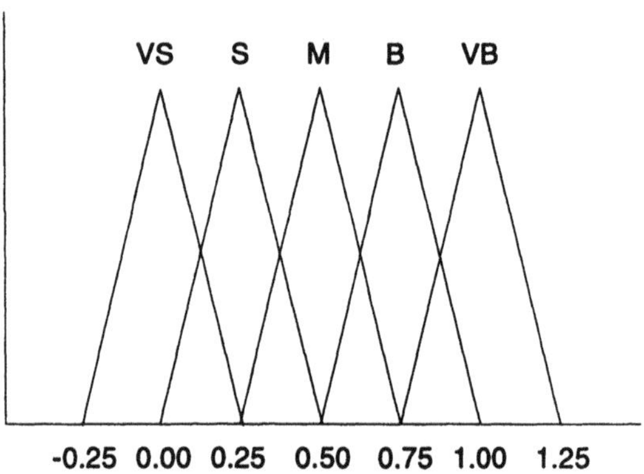

Fig. 13. Membership functions for the tuning variable α.

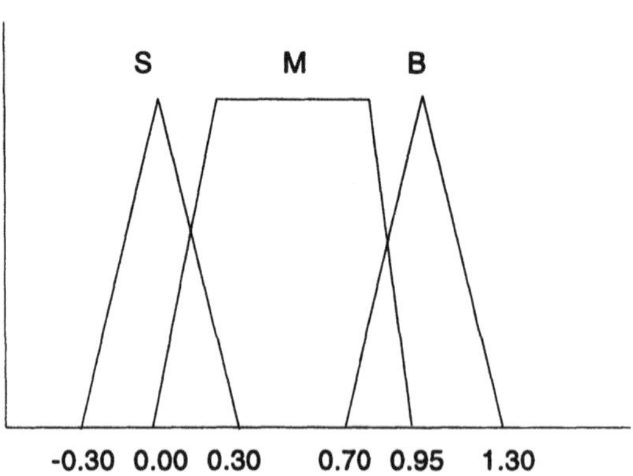

Fig. 14. Membership functions for the tuning variable β.

Step 4: Selection of a method of defuzzification. The centre of area method is used to transform the outputs of the fired rules into crisp outputs. After calculating the tuning parameters, new values for PI controller are calculated

$$K_c = 1.2\alpha K_u \qquad (7)$$

$$T_i = 0.75\frac{1}{1+\beta}P_u \qquad (8)$$

Parameters K_u and T_u are defined experimentally by setting the control loop in sustained oscillations and by measuring the response.

6 Simulation Results

The applicability of the fuzzy adaptive tuning methods presented by Mudi & Pal (1998), Chung et al.(1998) and Ramkumar & Chidambaram (1995) have been tested for the control of the output moisture of solids in the pilot plant dryer. The simulations are based on the mathematical model of the pilot plant dryer (Yliniemi 1999) implemented in Matlab functions. The proposed controllers have been constructed using Simulink and Fuzzy Logic Toolbox. Step changes to the input moisture of solids are made for examining the control behaviour of the dryer. The results achieved by non-adaptive and adaptive fuzzy and hybrid controllers where the FLC is combined with the PI controller are compared both qualitatively and quantitatively using time-integral performance indices IAE (integral of the absolute error) and ITAE (integral of time-weighted absolute error). The IAE uses the magnitude of the error, therefore both negative and positive variations cause the index to increase. The ITAE may be used to reduce large initial errors or to put an emphasis on the error occurring later in the response.

Figures 15 and 16 and Table 5 present the simulation results when the direct FLC, the self- tuned FLC, the hybrid PI-FLC and the self-tuned hybrid PI-FLC are used for controlling the output moisture of solids. The adaptive construction is based on the method of Mudi & Pal (1998)

Table 5. Performance indices for the different controllers.

Figure no°	Controller	IAE	ITAE
15	Direct FLC	0.1166	55.791
	Adaptive FLC	0.0810	31.816
16	Hybrid PI-FLC	0.0294	3.4739
	Adaptive hybrid PI-FLC	0.0203	1.6888

As Figures 15 and 16 and indices show, in both cases the adaptive controllers improve the performance decreasing the settling time (ITAE) and slightly decreasing the maximum error (IAE). The results also show that the hybrid controller gives less oscillation than the direct fuzzy logic controller.

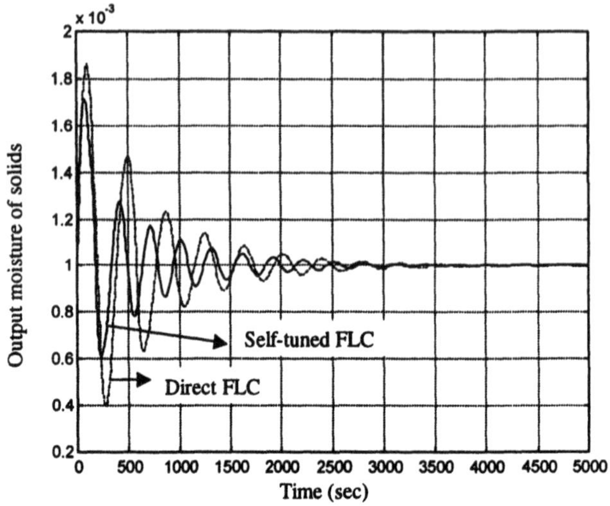

Fig 15. The behaviour of the output moisture of solids; the self-tuned direct FLC and the direct FLC for a step change in the input moisture of solids from 2.4 m-% to 3.4 m-%.

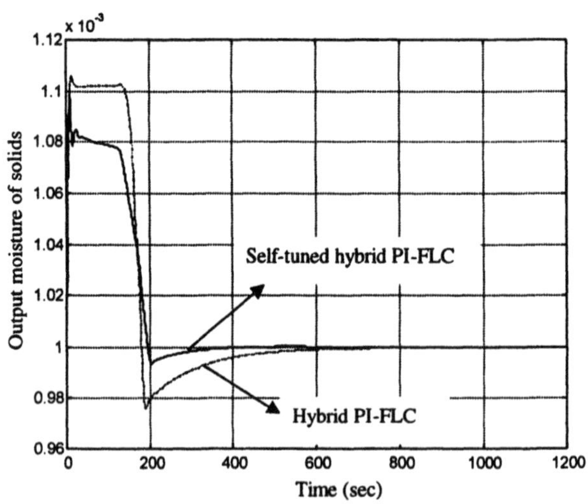

Fig 16. The behaviour of the output moisture of solids; the self-tuned hybrid PI-FLC and the hybrid PI-FLC for a step change in the input moisture of solids from 2.4 m- % to 3.4 m-%.

The other adaptive construction tested for the pilot plant rotary dryer is based on the method of Chung et al. (1998). For simulation studies the convergent coefficient value of w=3 was chosen as it gives the least oscillation, but the method did not improve the control performance.

The adaptive tuning based on Ramkumar & Chidambaram (1995) includes the ultimate gain and period, which are determined experimentally with step experiments. Figure 17 presents the sustained oscillation of the output moisture of solids with the PI control. The parameters of the PI controller are calculated from the equations 7 and 8. Simulation studies with different step changes on the input moisture of solids are made for determining the fuzzy ranges for α, β and for the error of the output moisture of solids. Fig. 18 and Table 6 present the control results achieved by the non-adaptive and fuzzy adaptive tuned PI controller. Fig. 19 shows one example of step responses gained with this method.

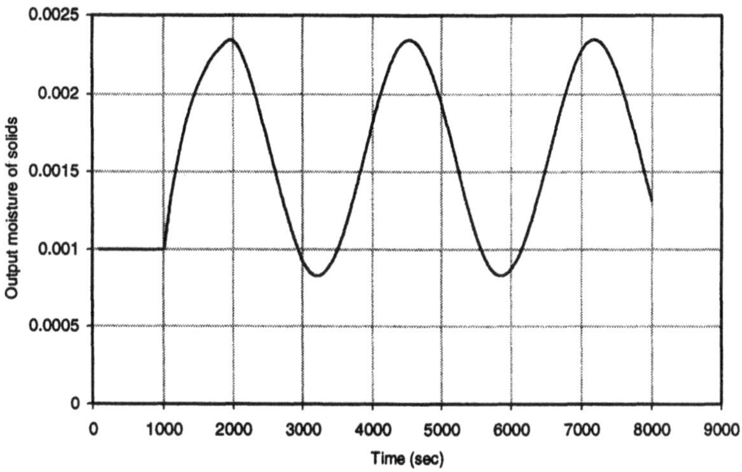

Fig. 17. Sustained oscillation of the output moisture of solids with the PI control.

Table 6. Performance indices for the different controllers.

Case	Controller	IAE	ITAE
1	Non adaptive PI controller	0.5634	1885
2	Fuzzy adaptive PI controller	0.4466	1242

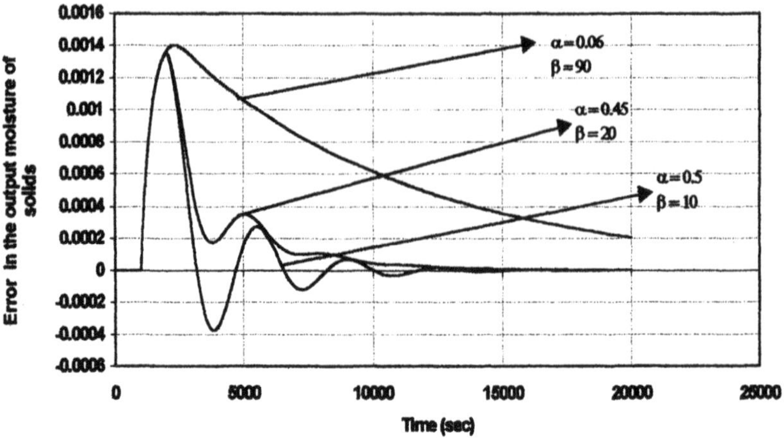

Fig. 18. Simulation results for determining the fuzzy ranges for α, β and error in the output moisture of solids.

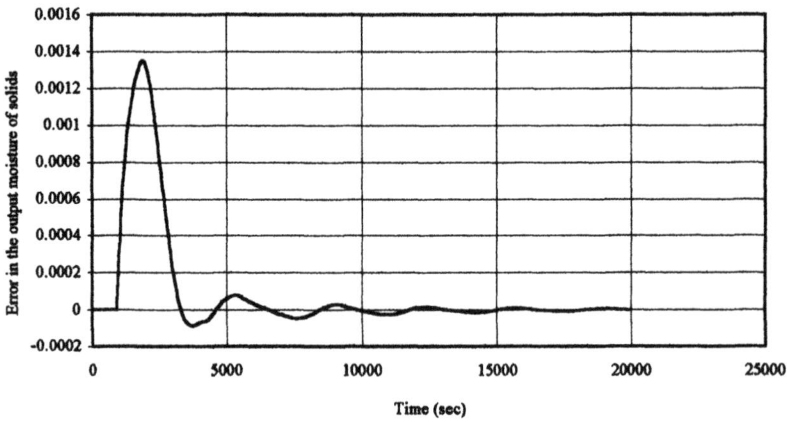

Fig. 19. The step response of the fuzzy adaptive PI controller.

7 Conclusions

This paper surveys the control applications of rotary dryers and the adaptive fuzzy methods. Three different methods were tested for the dryer control.

The first method was based on the previously applied hybrid PI-FLC controller with a simple gain adjusting approach. The output scaling factor is adjusted based on the experimentally defined rule base and membership functions. Good results were achieved and the controller performance was clearly improved.

The second method introduced a fuzzy PI controller and adjusting all scaling factors; for both inputs and the output. Results were not so encouraging, probably because of the fact that both the controller and the tuning part utilised rule bases and membership functions taken from the literature.

The third approach was a simple procedure for tuning normal PI controller with a fuzzy tuner. Once again, the adaptive controller performed better than the original one.

References

Baker, C. G J (1983): Cascading rotary dryers. In: Proc. Drying'83, 2, 1–48

Beck, M S, Bunn, P R, Gough, N E & Wormald C N (1971): Computer control of a pilot scale rotary solids drier. In: Preprints 3rd IFAC/IFIP Conference on Digital Computer Applications to Process Control, Helsinki, Part 2: XII-4, 1–7

Cammarata, L., Yliniemi, L. (1999): Development of a self-tuning fuzzy logic controller for a rotary dryer. Report A No 10, Control Engineering Laboratory, University of Oulu.

Chen, J.Y., Lin Y.H. (1997): A self-tuning fuzzy controller design. In: Proc. IEEE International Conference on Neural Networks 3, 1358–1362

Chiricozzi, E., Parasiliti, F., Ptursini, M., Zhang D.O. (1995); Fuzzy self-tuning PI control of PM synchronous motor drives. In: Proc. International Conference on Power Electronics and Drive Systems 2, 749–754

Chung, H.Y., Chen, B.C., Lin, C.C. (1998): A PI-type fuzzy controller with self-tuning scaling factors. Fuzzy Sets and Systems 93, 23–28

Daugherity, W.C., Rathakrishnan, B., Yen, Y. (1992): Performance evaluation of a self-tuning fuzzy controller. In: Proc. IEEE International Conference on Fuzzy Systems 389–397

Douglas, P.L., Kwade, A., Lee P.L., Mallick, S.K., Whaley, M.G. (1992): Modelling, simulation and control of rotary sugar dryers. In: Proc. Drying'92, 1928–1933

Duchesne, C., Thibault, J., Bazin, C. (1997): Dynamics and assessment of some control strategies of a simulated industrial rotary dryer. Drying Technology 15(2), 477–510

Harbert, F.C. (1973): Control of dryers by the temperature difference technique. Instruments and Control Systems 46 (9), 71–72

Harbert, F.C. (1974): Automatic control of drying processes: Moisture measurement and control by the temperature difference method. Chem. Eng. Sci. 29, b888–890

He, S.Z., Tan, S., Xu, F.L. (1993): Fuzzy self-tuning of PID controllers. Fuzzy Sets and Systems 56, 37–46

Jung, C.H., Ham, C.S. Lee, K.I. (1995): A real-time self-tuning fuzzy controller through scaling factor adjustment for the steam generator of NNP. Fuzzy Sets and Systems 7,: 53–60

Koskinen, J., Yliniemi, L., Leiviskä, K. (1998): Fuzzy modelling of a pilot plant rotary dryer. In: Proc. UKACC International Conference on CONTROL'98. Swansea, 1, 515–518

Koskinen, J. (1999): Rumpukuivaimen sumea mallintaminen. Lic.Thesis. University of Oulu, Department of Process Engineering

Lui, H.C., Gun, M.K., Goh, T.H. Wang, P.Z. (1994): A self-tuning adaptive resolution (STAR) fuzzy control algorithms. In: Proc.3rd IEEE World Congress on Computational Intelligence. 3, 1508–1513

Miyata, H., Ohki, M., Ohikita, M. (1996): Self-tuning of fuzzy reasoning by the steepest descent method and its application to a parallel parking. IEICE Transactions on Information and Systems e79-d (5), 561–569

Mudi, R.K. Pal, N.R. (1998): A robust self-tuning scheme of PI-type and PD-type fuzzy controllers. IEEE transactions on Fuzzy Systems 7(1), 2–16

Perry, J.H. (1963): Chemical Engineers' Handbook . McGraw-Hill Inc, New York, 22–1–22–107

Pietranski, J.F., Marsolan, N.F. King K.-H. (1982): Expert fuzzy process control of a rotary dryer. In: Proc. American Control Conference, New York, 1359–1362

Ramkumar, K.B., Chidambaram, M. (1995): Fuzzy self-tuning PI controller for bioreactors. Bioprocess Engineering 12(5), 263–267

Robinson, J.W. (1989): The Delta T-A new drying model for pulp and paper. In: Proc. TAPPI Engineering Conference. Tappi Press, Atlanta, 183-187

Robinson, J.W. (1992): Improve dryer control. Chem. Eng. Progress, December 28–33

Shinskey, F.G. (1974): Process control systems with variable structure. Control Engineering, August 63–66

Strumillo, C. Kudra, T. (1986): Drying: Principles, Applications and Design. Gordon and Breach Science Publishers, Montreaux

Wright, D.J. (1976): Continuous moisture measurement in mining processes. In: Proc. Instrumentation in the Mining and Metallurgy Industries, Vancouver, 8–10

Yliniemi, L., Koskinen, J., Leiviskä, K. (1998): Advanced control of a rotary dryer. In: Heidepriem, J. (ed.): Automation in Mining, Mineral and Metal Processing Preprints. Elsevier Science, NewYork, pp.127–132

Yliniemi, L. (1999): Advanced control of a rotary dryer. Dissertation, University of Oulu, Department of Process Engineering

Zheng, J., Guo, P.,.Wang, J.D. (1992): STFC – self-tuning fuzzy controller. In: Proc. IEEE International Conference on Systems, Man and Cybernetics 2, 1603–1608

Soft Computing at a Flotation Plant

Heikki Hyötyniemi[1], Raimo Ylinen[2], Jorma Miettunen[3]
[1]Helsinki University of Technology, Control Engineering Laboratory, PL 5400, FIN-02015 TKK, Finland
[2]University of Oulu, Systems Engineering Laboratory, P.O.Box 4300, FIN-90014 Oulun yliopisto, Oulu, Finland
[3]Outokumpu Mining Oy, Pyhäsalmi Mine, Outokumpu Mining Oy, Pyhäsalmi Mine, P.O. Box 51, 86801 Pyhäsalmi, Finland

1 Introduction

Flotation is used in mineral processing industries for separation of grains of valuable minerals from those of side minerals (Laskowski and Woodburn 1998). In the continuous flow flotation cell (Fig. 1), air is pumped into a suspension of ore and water. The desired mineral tends to adhere to air bubbles and rises to the froth layer where the concentrate floats over the edge of the cell; the main part of other minerals remains in the slurry. The separation of minerals requires that the desired mineral is water-repellent: in zinc flotation, this can be reached by conditioning chemicals as copper sulphate $CuSO_4$; xanthate is needed to reach lower surface tension, etc.

An industrial scale flotation plant consists of complicated flotation circuits where several flotation cells are connected in series and in feedback loops (Fig. 2). The rougher and scavenger sections are responsible for obtaining recovery, while concentrate grade is attained in the cleaner circuit. The fundamental problem in flotation is that a good recovery and a high concentrate grade are conflicting objectives. The optimum operation point is depending on the actual economic situation.

The process can be controlled by chemical variables like reagent additions and addition points, or by physical variables, e.g. air feed, pulp flows, and collection points for concentrate. Many disturbances affect the process, for example feed grade, fineness of crystallisation, degree of oxidation, fineness of grind, pulp density and volumetric flow rates, the last three of which can also be manipulated to some degree. The output variables can be divided into intermediate variables as grade, recovery, pulp density, flow rate, froth level, etc. of rougher and scavenger concentrates and performance variables, which are the corresponding variables of the final concentrate. For each cell, there are also several internal state variables describing e.g. the concentrations of the chemical and physical components of the pulp and froth in the cell.

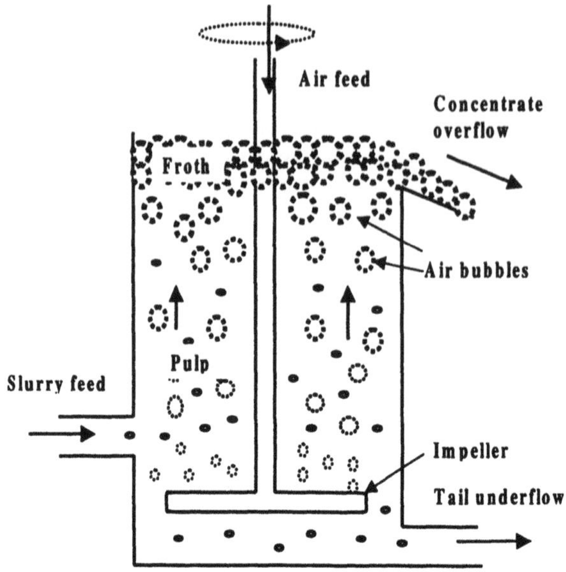

Fig. 1. Flotation cell.

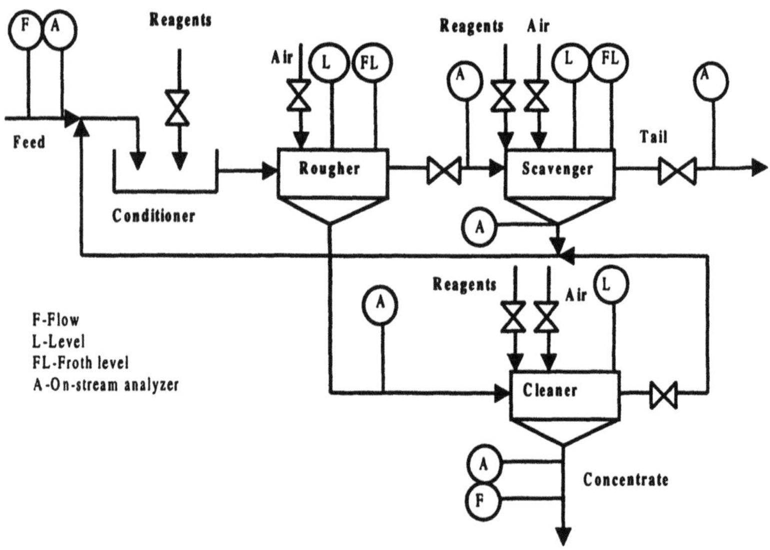

Fig. 2. Typical flotation circuit with instrumentation.

Flotation is one of the most difficult and challenging processes in mineral processing industry. This characteristic of the process mainly arises from the inherently chaotic nature of the underlying microscopic phenomena; there are no good models available that would capture the behaviour of the particles. Additional problems are caused by the fact that today's measurement technology is not able to provide with a description of the current state of the process that would be accurate and reliable enough. It is the froth surface that dictates the quality of the out flowing concentrate; the properties of the froth are reflected in its texture, movement, and colour. No standard measurement devices, however, can capture the outlook of the froth.

Thus, human operators control the most of the chemical reagents that are used to increase the efficiency of flotation. The operators usually determine the suitable levels of the reagents by analysing the visual appearance of the froth; the control strategies they apply consist of expert knowledge.

The fast development of information technologies, however, has made it possible to acquire images of the froth in real-time and automatically extract features from the froth image that resemble the features used by the operators. Thus, the limited capacity of the operator to monitor cells continuously (the operator is usually responsible for various circuits consisting of several cells) can be increased by installing a video camera over critical cells and connecting the camera to a computer that is able to process grabbed images in real-time. The importance of this topic is illustrated by the fact that it has been studied extensively in the mineral engineering community. A large number of different research papers have been published on the subject (for example, see (Cipriano et al. 1998; Moolman et al. 1995a, b; Niemi et al. 1997)).

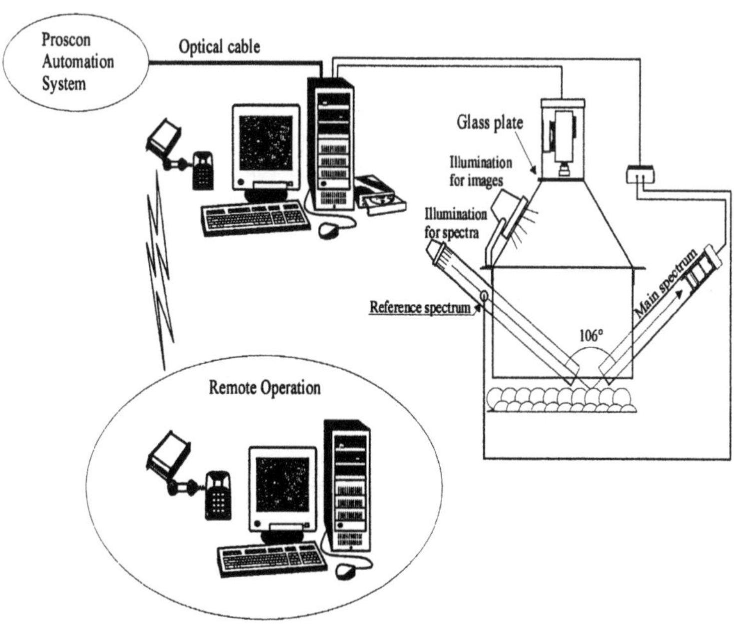

Fig. 3. The froth analysis system (courtesy of Mr. Jani Kaartinen).

This paper reports the results achieved during the ESPRIT LTR project ChaCo (Characterization of Flotation Froth Structure and Colour by Machine Vision); see (Ylinen et al. 1999). Fig. 3 illustrates the overall structure of the system developed during the project.

2 Soft Computing Techniques for Flotation

To realise reliable and efficient operation of an on-line flotation plant, there are various tasks that need to be automated:

- The relevant process variables have to be estimated using machine vision
- The state of the process has to be detected
- Appropriate control actions need to be chosen.

All these tasks require "applied AI techniques", and each of these tasks is discussed separately in what follows.

2.1 Soft Sensors Based on Image Analysis

The status of the flotation froth cannot be uniquely defined; there are just a few measurements that can be explicitly defined and determined (like the froth level and thickness), whereas most of the factors that characterise the froth properties are more or less conceptual having no explicit definition. However, as compared to many other image analysis tasks, now the pattern recognition task is rather simple; the structure of the images is known before hand, and only statistical features need to be extracted.

To construct "soft sensors" for the non-measurable quantities, operator interviews were first carried out to find out what are the most relevant phenomena, and mathematical interpretations for them were derived (Hätönen 1999). It turned out that the static image alone is not enough; it is the dynamics of the froth layer that tells very much about the process state.

Typical "dynamical" variables extracted from the collected froth images are (average) speed, speed variation, and bubble collapse rate. The speed of the froth is obtained by computing the cross-correlation matrix (by using the 2-dimensional Fourier transformation) between two images (captured using 0.2 sec time interval) and determining its maximum value; the speed of the froth is given by the location of the maximum peak. The value of the maximum peak, on the other hand, gives a simple measure for the speed variation (homogenous movement of the froth results higher peak in the cross-correlation matrix). The number of pixels in which the difference between two consecutive aligned images is larger than some given threshold, is used as a measure for the bubble collapse rate (BCR).

The colour of the froth is also of interest; colour variables are computed in two spaces: RGB (red, green, blue) and HSV (hue, saturation, value). The distributions of colour values are characterised by their mean, standard deviation, skewness and curtosis. Skewness represents the symmetricity of the distribution, and curtosis stands for its "peakedness".

Examples of extracted structural features are area, aspect ratio, perimeter, roundness and transparency of bubbles. Aspect ratio is the ratio between the major axis and the minor axis of the ellipse equivalent to the bubble; perimeter stands for the length of the outline of a bubble, and roundness of a bubble is determined by the formula roundness = $(\text{perimeter})^2 / (4\pi \times \text{area})$. Transparency is measured by calculating the points with especially high reflection – light reflected from bubbles that are covered with concentrate is rather scattered. To calculate these features the bubble images have to be segmented and for segmentation, the watershed technique was applied; see (Beucher and Mayer 1993). After the segmentation, the bubbles can be considered as separate objects and the features that are of interest and also their statistical distributions can be calculated.

Further, the extracted features were applied for the extraction of more advanced information like mineral concentrations. The parameters of the models were tuned using statistical multivariate methods (Basilevsky 1994). For instance, principal component analysis (PCA) and partial least squares estimation (PLS) have been applied (Hätönen et al. 1999a, b). It was recognised that red colour intensity correlates strongly with the zinc concentration in the froth; the spatial variance of the froth speed also correlates positively with zinc content, whereas BCR correlates negatively.

2.2 Categorisation of Froth Layers

When machine vision has been applied to analysis of froth surfaces, normally the froth classes are labelled using rather low-level features – for example, there may be froths with big or small bubbles, etc. (Moolman et al. 1995b). Now, on the other hand, operator interviews revealed that the most interesting (fuzzy) froth categories in zinc flotation are wet, dry, and stiff. Loosely speaking these froth types can be characterised so that wet froth typically has "empty bubbles" (the concentrate load being low); dry froth has "hexagonal bubbles" (the bubble tessellation being rather evenly distributed); stiff froth is "porridge-like". These categories are conceptual and there are no exact mathematical definitions for them; however, many of the extracted low-level features (see previous section) correlate with these categories. For example, for wet froth the transparency value and bubble collapse rate are high, for dry froth the bubble size distribution is even, and for stiff froth the speed and speed variance are low. Using the multivariate data analysis terminology, the froth prototype vectors can be seen as (non-orthogonal) latent variables underlying the measured variables.

For the classification of the froth three different methods were implemented in the analyser prototype. The first was the so called simple froth classifier, which measures the similarity of a given froth image against the predefined froth prototypes. This has been also realised using fuzzy logic. The second method was a tree-classifier, and the third classification method used the so called GGHA method to analyse the statistical structure of the data (Hyötyniemi and Hasu 2000).

Whereas the froth classification is not used directly for control (see next section), its role is important in process monitoring and supervision. The optimal froth is rather dry; bubbles are loaded with concentrate, but the froth still floats out from the cell. The most critical case is the extremely stiff froth: This pathological situation may result in the collapse of the froth surface, so that no concentrate flows out and it takes typically 10–20 minutes before normal production can be recovered (see Figs. 4 and 5). It turned out that the implemented classifiers were capable of detecting the risk of collapse some 10 minutes before the surface finally collapsed, so that some corrective actions could still be applied (Hyötyniemi and Ylinen 1998).

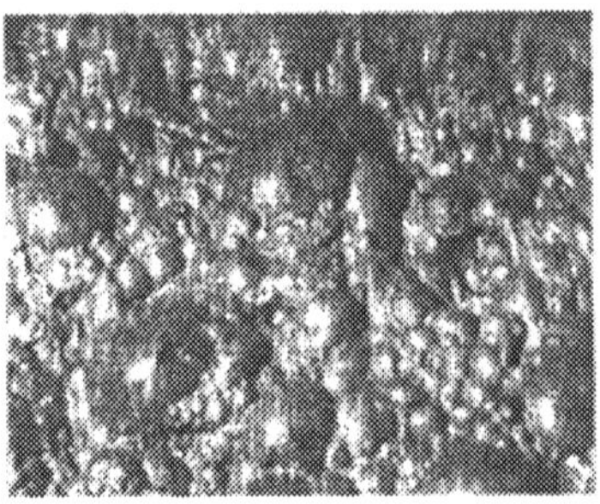

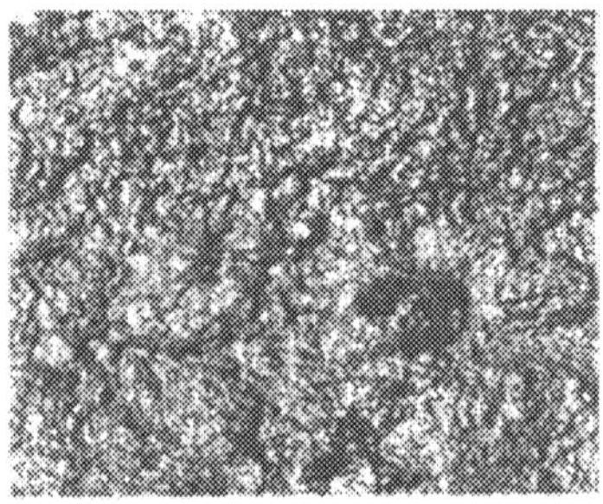

Figs. 4 and 5. On the left, a typical "living" froth surface, and on the right, a collapsed one

2.3 Expert Control of the Flotation Process

The image analyser described above was first installed at the first rougher cell of the zinc flotation circuit in the Pyhäsalmi Mine of the Outokumpu Mining Company. Fig. 6 presents the flowsheet of the corresponding part of the process.

The image analyser gives a number of variables, which can be used for control. Based on the literature study, operators' knowledge, and on experimental work, the following five variables seemed to be the most important: bubble collapse rate, transparency of bubbles, bubble size, red colour intensity, and froth speed. Fig. 7 shows how one of these image variables (bubble collapse rate) affects one of the process variables, the zinc recovery rate. BCR may be a key variable because it seems to be possible to define an area where it should be maintained in order to guarantee good zinc recovery (the minimum in the graph). The optimum area of BCR, however, moves with the zinc content of the flotation feed, as can be seen in Fig. 8.

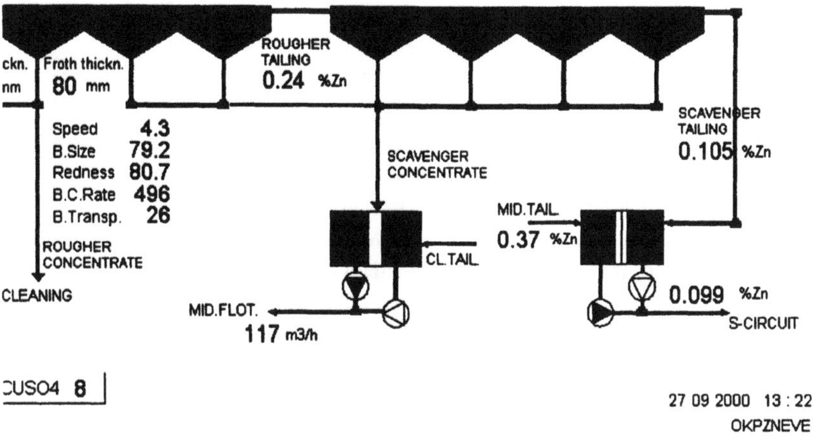

Fig. 6. Rougher and scavenger cells of the Pyhäsalmi zinc circuit.

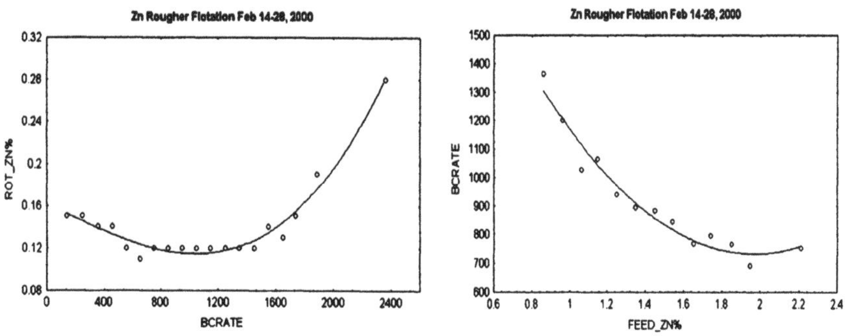

Figs. 7 and 8. On the left, the Zn content in the rougher tailing (that is, the amount of zinc lost in the first phase of flotation) is shown as a function of bubble collapse rate – apparently, an optimum exists; however, as shown in the figure on the right, controlling the bubble collapse rate is not straightforward, as it depends on various uncontrollable factors, like the incoming Zn grade.

Fig. 9 shows the relationship between BCR and the feed of copper sulphate. The figure clearly indicates that BCR can be controlled with copper sulphate. In reality this relationship is more complex because both variables depend more or less on the zinc grade of the feed; however, added copper sulphate should always decrease BCR. It seemed that bubble transparency is also a good indication of the flotation performance.

Overdosage of copper sulphate is indicated by a very low value of bubble transparency (Fig. 10). On the other hand, a high transparency value means the lack of copper sulphate. Like BCR, the bubble transparency value also depends on the zinc grade of the feed, but the optimum area is not quite constant.

Based on these experiences, a rule-based control algorithm for the dosage of copper sulphate was constructed. The (simplified) rule set can be presented as shown in Table 1. The action part of the rules has a very simple form: the flow of $CuSO_4$ is either increased ("+") or decreased ("–") by a fixed amount (about 2% of the operating range at a time). As compared to normal expert systems, now the control flow is fixed and there is a ranking between the rules. If a rule with higher priority can be applied, it is always selected instead of the lower priority ones.

The knowledge in the rule base combines the observed relationships within the process with the objectives of production. The controller is non-model-based (there is no dynamic description of the influence mechanisms), only the (assumed) causal dependencies are applied. The basic assumption is that increasing $CuSO_4$ makes more zinc adhere to the bubbles, making them more heavily loaded.

The most critical rules with the highest ranking try to avoid the froth collapse; that is, if the bubbles are too loaded, the dosage of $CuSO_4$ is made smaller. If rule #1 is active (the froth thickness is below some preset value) this is already too late, whereas rule #2 may help in avoiding the collapse. The next rules (#3 and #4) try to minimise the amount of zinc passing through the system and being lost. The next rule #5 prevents the excessive foam. Next, in the rule #6, if any of the image variables indicate too wet bubbles, more $CuSO_4$ is pumped in. Finally, if everything works all right, so that none of the previous rules was active, the consumption of $CuSO_4$ is minimised.

Fig. 11 shows one week's period of the copper sulphate control. It can be noticed that the control stabilised the system and no collapse situations existed.

Table 1. The $CuSO_4$ control rule base that mimics the operator actions

Ranking	Condition	Action
1.	IF froth thickness < lower limit	–
2.	IF BCR < lower limit OR bubble transparency < lower limit	–
3.	IF zinc content in rougher tailing > upper limit	+
4.	IF zinc content in scavenger tailing > upper limit	+
5.	IF froth thickness > upper limit	+
6.	IF BCR OR bubble transparency OR bubble size>upper limit	+
7.	ELSE	–

Process tests with xanthate showed that red colour and bubble size were reacting to the set point changes of xanthate. It also seemed that increasing xanthate consumption would replace part of more expensive copper sulphate while the flotation results remained the same or even improved. Therefore, a control logic including red colour and bubble size was built up.

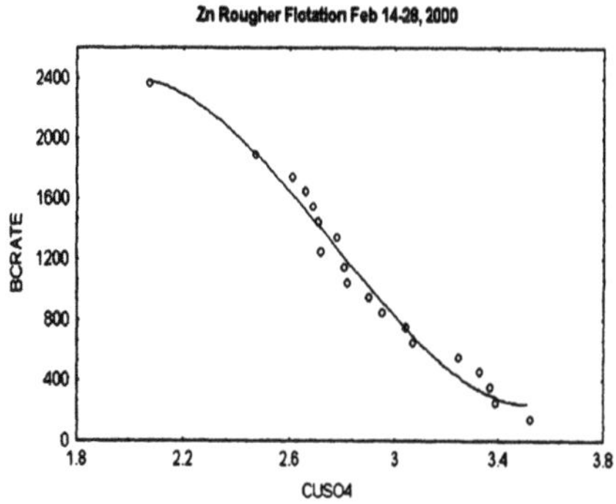

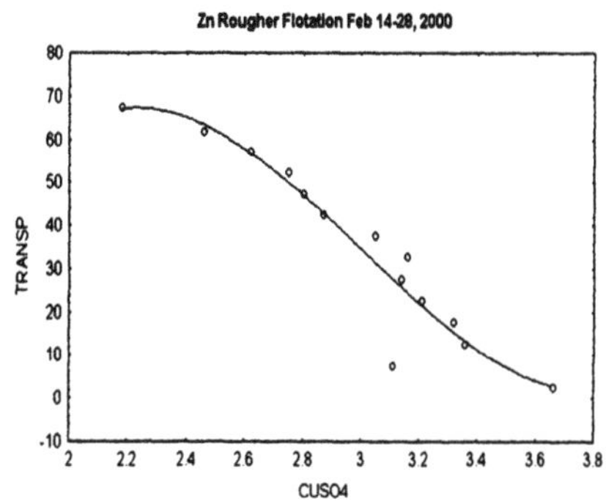

Figs. 9 and 10. In these figures, the effect of copper sulphate is shown; as $CuSO_4$ increases the bubble collapse rate decreases (the bubbles become stiffer). This applies also the bubble transparency (the bubbles are more "heavily loaded")

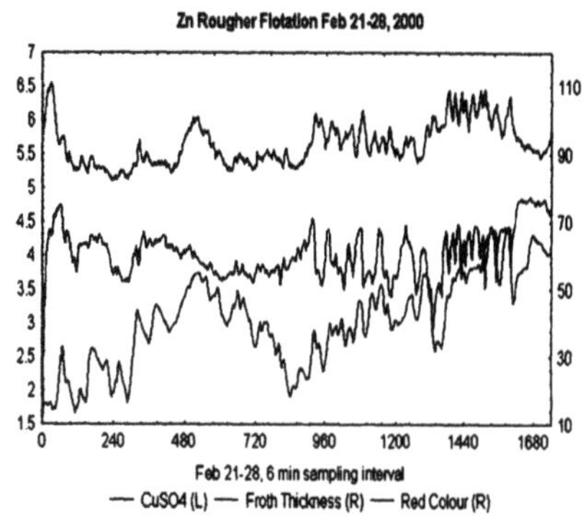

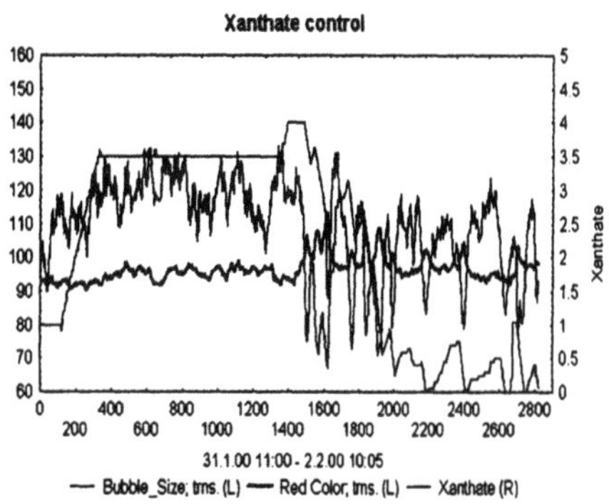

Figs. 11 and 12. Real-time control using the "expert controllers". On the left, the feed of $CuSO_4$ is shown (lowest curve) against the measured froth thickness and red colour channel intensity (center and top, respectively); on the right, xanthate feed is shown against bubble size and red colour intensity.

Fig. 12 shows results of a test run. In this test the xanthate set point has been on the high alarm limit for the first half of the two-day period because big bubble size suggested that there is a lack of xanthate. During the other half of the period the control was working properly. The reason of the poor control seems to be a low zinc grade in the feed proposing that the bubble size limit should be a function of the feed grade.

3
Conclusions

During the project, the machine vision set-up for permanent recordings at the flotation plant was designed and built. RGB camera, image grabber card, spectroradiometer, and the PC computer equipment were installed at the Pyhäsalmi flotation plant in the first rougher cell of the zinc circuit. The camera was installed inside a metal hood to protect it against dirt. The geometrical shape of the hood was selected so that the homogenous illumination of the froth would be obtained for an easy processing of the images. The user interface was implemented in the LabVIEW© environment, and the on-line analysis system was connected to the PROSCON© process automation system.

In July 1999, the froth analyser was connected in the local network of Pyhäsalmi concentrator plant. The hardware and software have been operating rather reliably since then. In the spring 2000, the analysis system was integrated in the closed loop control system. Because of the natural variations in the incoming ore, unambiguous comparisons against the earlier control performance cannot be made; however, according to visual inspections the results seem promising. Simply being able to avoid the froth collapse situations in advance saves a lot of raw material, time, and money. Perhaps the added value of the new approach is best illustrated by the fact that another identical analysis system is being installed in a separate flotation circuit in Pyhäsalmi. Similar system is being experimented also at the Boliden Garpenberg plant in Sweden (Boliden mining company was the other industrial partner in the ChaCo project).

Even though the industrial application seems to be promising, something seems to be missing from the AI point of view: the somewhat obscure feel of "intelligence" vanishes when explicit algorithms can be written for implementing all specific tasks. Expert systems boil down to a set of if–then rules, pattern recognition to finding measurements characterising the process state, and categorisation to measuring distances between data clusters. However, after all, the cumulation of complexity in the overall system does the trick: seen from outside, the consistent operation of the controlled plant still looks rather intelligent.

Acknowledgement

This research was carried out as a part of the EU funded project "Characterization of Flotation Froth Structure and Color by Machine Vision (ChaCo)".

References

Basilevsky, A. (1994): Statistical factor analysis and related methods. John Wiley & Sons, New York

Beucher, S. and Meyer, F. (1993): The morphological approach to segmentation: The watershed transformation. In: Dougherty, E.R. (ed.): Mathematical Morphology and Image Processing. Marcel Dekker, New York, pp. 433–481

Cipriano, A., Guarini, M., Vidal, R., Soto, A., Sepúldeva, C., Mery, D., Briseño, H. (1998): A real time visual sensor for supervision of flotation cells. Mineral Engineering 11(6), 489–499

Hyötyniemi, H., Ylinen, R. (1998): Modeling of visual froth data. In: Heidepriem, J. (ed.): Preprints of the IFAC Symposium on Automation in Mining, Mineral and Metal Processing. Elsevier Science, New York, pp. 309–314.

Hyötyniemi, H., Hasu, V. (2000): Independent components of expertise. In: Hyötyniemi, H. (ed.): Proceedings of the Finnish Artificial Intelligence Conference (SteP 2000). Finnish Artificial Intelligence Society, Helsinki, vol. 3, pp. 37–44

Hätönen, J. (1999): Image analysis in mineral flotation. M.Sc. Thesis, Helsinki University of Technology, Control Engineering Laboratory, Report 116

Hätönen, J., Hyötyniemi, H., Bonifaci, G., Volpe, F., Serranti, S., Carlsson, L.-E. (1999): Using PCA in controller strategy design for a flotation process. In: Chen, H.F., Cheng, D.Z., Zhang, J.F. (ed.): Preprints the 14th IFAC World Congress. Elsevier Science, Oxford New York Tokyo, pp. 385–390

Hätönen, J., Hyötyniemi, H., Miettunen, J., Carlsson, L.-E. (1999): Using image information and PLS for predicting mineral concentrations in the flotation froth. In: Proceedings of the Second International Conference on Intelligent Processing and Manufacturing of Materials (IPMM'99), Hawaii, USA

Laskowski, J. S., Woodburn, E. T. (eds.) (1998): Frothing in flotation II. Gordon and Breach Science Publishers

Moolman, D. W., Aldrich, C., van Deventer, J. S. J., Bradshaw, D. B. (1996): The interpretation of flotation froth surfaces by using digital image analysis and neural networks. Chem. Eng. Sci. 50(22), 3501–3513

Moolman, D. W., Aldrich, C., van Deventer, J. S. J., Stange, W. W. (1995): The classification of froth structures in copper flotation by means of a neural net. Int. J. Mineral Processing 43 (3–4), 193-208

Niemi, A., Ylinen, R., Hyötyniemi, H. (1997): On the characterization of pulp and froth in cells of flotation plant. Int. J. Mineral Processing 45(3), 51, 51–66

Ylinen, R., Hasu, V., Hyötyniemi, H., Sipari P. (1999): control of flotation by machine vision. In; Proceedings of the Automation Days 1999. Finnish Society of Automation, Helsinki, pp. 459–464

Part III

Paper Industries

Intelligent Modelling of Continuous Pulp Cooking

Kauko Leiviskä, Esko Juuso, Ari Isokangas
University of Oulu, Control Engineering Laboratory, P.O.Box 4300, FIN-90014 Oulun yliopisto, Oulu, Finland

1 Introduction

Cooking is the major process in the pulp mill and its proper control is very important to the pulp production. The cooking should be done to the best quality at the minimum cost. Usually, the best quality means the minimisation of variation in the quality variables. Quality variations must simultaneously compensate for the effects of several types of disturbances. The main objective is to guarantee the similar cooking history for all chip particles as they pass through the digester. This leads to decreased variation in pulp quality compared with the conventional control.

Pulp cooking has used computerised control since the early 1960s (Sutinen et al. 1990; Leiviskä 1999). Chip quality variations, measurement problems and long process delays make the control difficult. The main problem is that the main quality variable, the Kappa number, can not be measured while the cooking goes on. Kappa number tells the extent of the delignification reaction and correlates with the hardness of pulp.

CLA 2000 Cooking Liquor Analyser developed by ABB gives new possibilities to monitor the proceeding of cooking. The concentrations of alkali, total dissolved solids and lignin are measured on-line during individual cooks in a batch digester house and from several circulations in a continuous digester. Using these measurements it is possible to predict Kappa number before the end of the cook in the batch digester or to control cooking conditions in the continuous digester to obtain the desired Kappa number. These cooking concentration measurements give also other valuable information.

With these measurements it is possible to estimate the development of the Kappa number long before the end of the cook. The digester process is far from linear and simple input-output system. With these basics in mind a tool for analysing the measurements has to be able to deal with non-linear modelling and multivariable inputs.

Different approaches have been used for mathematical modelling of the cooking result (Haataja et al. 1997; Lemmetti et al. 1998; Murtovaara et al. 1998a, b, 1999): fuzzy logic (FL), partial least squares method (PLS), artificial neural networks (ANN) and linguistic equations (LE).

2 Process Description

2.1 Continuous Cooking

Chemical pulping includes all methods of pulp manufacturing in which raw materials (usually wood) containing cellulose fibres are treated with aqueous alkaline or acidic solutions (cooking liquor). The aim of chemical pulping is to remove enough lignin from wood so that the fibres are free and give them the required characteristics at the lowest possible cost. The main process nowadays is the sulphate pulping process.

In sulphate pulping the cooking liquor is alkaline. The active chemicals are sodium hydroxide (NaOH) and sodium sulphide (Na_2S). The process is named after sodium sulphate (Na_2SO_4), which is used as a make-up chemical to replace sodium and sulphur losses.

There exist two main types of process equipment for cooking: batch and continuous digesters (see Fig. 1). This paper concerns only with the continuous cooking process.

In continuous cooking, chips and cooking liquor are fed into the top of the digester as shown in Fig. 1. The chips are impregnated with cooking liquor at the top of the digester, heated under the pressure and held at the certain cooking temperature for some time. In many cases a separate impregnation vessel is used. Cooking temperature depends on wood species and grade requirements and is between 150–170 °C. In the continuous digester, the cooking time depends on the production rate.

Cooking reactions start when chips reach the cooking temperature. The active chemicals in the cooking liquor react with lignin in chips and convert it chemically into the compounds that dissolve in the cooking liquor. Lignin is the bonding material in chips that holds individual fibres together. After dissolving the lignin, fibres are easily separated into the mass that is leaving the digester as pulp.

Cooking chemicals also react with fibrous material (cellulose and hemicelluloses) and break down their molecular bonds. These reactions are to be avoided as much as possible because they lead to losses in the fibre yield and decrease the pulp strength.

Pulp washing is started in the lower part of the digester as the counter current process. The liquor washed out from pulp is pumped to the evaporation plant. After the digester, pulp is further washed in separate processes, screened, cleaned and dried or pumped over to the paper mill.

Two main technological solutions exist for continuous cooking: hydraulic digesters and vapour phase digesters. Several modifications as Modified Continuous Cooking (MCC), Extended Modified Continuous Cooking (EMCC), and Iso-Thermal Cooking (ITC) have been referred to in Lemmetti et al. (1998).

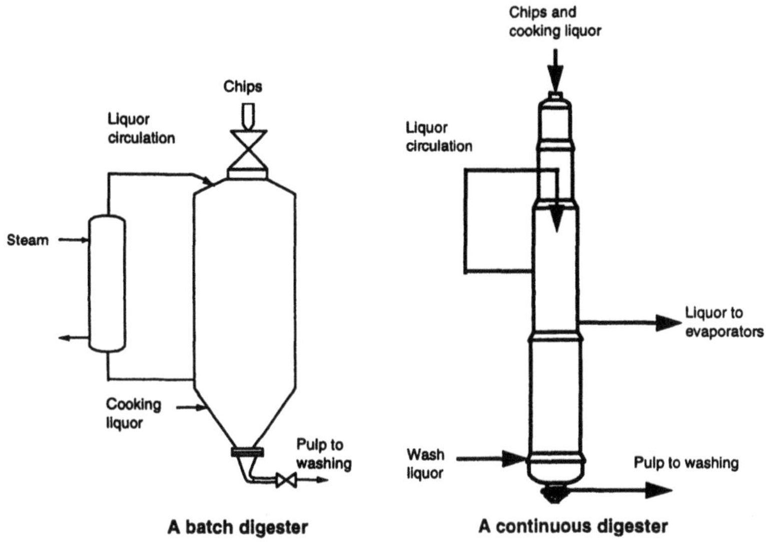

Fig. 1. Batch and continuous digesters.

2.2 Measurements

Measurements that are traditionally used in the sulphate cooking process are temperatures at the top of the digester and in cooking circulations and sometimes alkali measurements of the incoming liquor and cooking circulations. The liquor-to-wood ratio of the process is estimated from the measured amount of chips and their moisture content. The final cooking result (Kappa number) is measured from the blow line after the digester by using manual sampling and laboratory analysis or an automatic on-line device. So, with existing measurements long delays make it difficult to use feedback or feedforward control to change process conditions for stabilising the Kappa number.

Cooking liquor analyser CLA 2000 is an advanced measurement device developed for analysing chemical pulping processes (Sutinen 1996). It utilises a continuous sampling technique that means also, in practice, a continuous measurement if the analyser is applied for one sampling point, only. Cooking liquor analyser enables the analysis of alkali concentration and concentrations of total dissolved solids and dissolved lignin during individual batches in a batch digester house or from several cooking circulations in a continuous digester.

Measurements are carried out in the actual process concentration and no dilution is needed. Alkali is measured based on conductivity, total solids measurement is based on the refractive index, and dissolved lignin is measured using UV-absorption at process concentrations.

In the sulphate cooking on-line information given by the cooking liquor analyser improves the monitoring and control of the cooking process. The analysis of alkali and total dissolved solids at the beginning of the cook describes the true alkali charge and takes the most important factors in chip or liquor quality into account. As the cooking progresses, all three measurements give information showing how cooking reactions go on. With batch digester operation it is possible to terminate the cooking process just when the required cooking degree has been achieved, or to alter the final cooking conditions in continuous digesters to obtain the desired quality. Further possibility is to use the information as estimating yield and strength properties of the pulp.

2.3 Digester Control

Basic control functions in continuous digesters are given in Leiviskä (1999) and will not be repeated here. Only quality control is reviewed as follows. In the continuous digester, quality control faces with three main difficulties:

- Variations in chip quality
- Measurement problems
- Long process delays.

Control of pulp quality goes traditionally back to Kappa number control and further to controllable alkali and temperature profiles. The retention time is another crucial variable, but in the continuous digester it is determined by the production rate. Kappa number control tries to provide constant conditions for cooking reactions. Because of long retention times and analysing delay, a direct feedback from laboratory Kappa number is impractical. The blow line Kappa sensor decreases the analysing delay and allows the control interval decrease. By measuring liquor concentrations from different liquor circulations one can estimate Kappa number in different stages of the cooking process. With a good Kappa number estimate, control actions are possible before pulp reaches the blow line of the digester. This paper concerns with using intelligent methods in the Kappa number estimation from liquor concentrations

3 Process Data

The modelling was based on actual production data collected during two periods from a continuous digester house in Finland. It included the extraction flow (liquor flow to the evaporators in Fig. 1) measurements (conductivity (ExtrJK) and refractive index (ExtrRI)), temperature in the extraction flow (TExtr), and the measurement of Kappa number (Kappa) from an on-line device after the digester.

The first period was between February and May in 1997. The data (1263 measurements) was collected from the different time periods, when the production rate was around 900 tons a day. Some corrections to the data were done. If successive measurements deviated more than twice from the standard deviation, they were replaced by the mean values of two earlier and two following measurements. Statistical properties of data are shown in Table 1 and the correlation matrix is in Table 2. The correlation between Kappa and ExtrRI (-0.32), and between Kappa and ExtrJK (0.06) is not good, but the earlier experience has shown that it is still reasonable to use measurements from extraction flow (RI, JK and TE) in the modelling. Before the modelling the regression analysis was done for the selected variables (see equation 1). Correlation coefficient (R^2) was 0.40 and the standard error of estimate 1.2. All variables were statistically significant. Then data was divided into training and testing data (600 measurements each).

Table 1. Statistical parameters.

	ExtrRI	ExtrJK	Textr	Kappa
Average	17.51	2.23	167.49	29.70
Std. Dev.	0.25	0.04	0.51	1.55
Range	1.51	0.24	3.20	9.90
Minimum	16.70	2.12	165.40	25.70
Maximum	18.20	2.36	168.60	35.60

Table2. Correlation matrix.

	ExtrRI	ExtrJK	Textr	Kappa
ExtrRI	1.000			
ExtrJK	-0.016	1.000		
Textr	0.247	-0.256	1.000	
Kappa	-0.318	0.064	-0.590	1.000

Table 3. Statistical parameters for the second data set.

	ExtrRI	ExtrJK	Textr	Kappa
Average	17.39	2.23	166.93	30.30
Std. Dev.	0.23	0.02	0.45	0.89
Range	0.87	0.16	2.3	4.50
Minimum	17.00	2.15	165.9	28.00
Maximum	17.9	2.32	168.3	32.50

Linear regression equation for Kappa was

$$Kappa = 342.5 - 1.04*ExtrRI - 2.92*ExtrJK - 1.72*TExtr \quad (1)$$

Another data set was collected during one week's period when the production rate was about 1170 tons a day. There were 320 measurements available, and their statistical parameters are shown in Table 3.

4 Results

4.1 Comparison of Fuzzy Modelling Methods

The starting point for the comparison is the modelling work done using linguistic equations (Murtovaara et al, 1999). In linguistic equation approach, interaction matrices represent knowledge (rules). In this way, a symmetrical set of linguistic relations can be converted to a compact equation

$$\sum_{j=1}^{m} A_{ij} X_j = 0, \qquad (2)$$

where X_j is a linguistic level for the variable j, j=1...m. In the original system (Juuso and Leiviskä 1995; Juuso 1996), the linguistic values very low, low, normal, high, and very high were replaced by integer numbers -2, -1, 0, 1 and 2. The direction of the interaction is represented by coefficients A_{ij}, and originally there were only three different coefficients {-1,0,1}.

The original equation model has been extended to various directions. Different interaction strengths were introduced by using also other positive and negative integers as coefficients. This approach corresponds to handling different partitions in fuzzy set systems.

The tuning phase becomes easier when real-valued linguistic levels are used as the system is in a continuous form. This is especially useful when the system includes a large number of variables. Real-valued approach is now the main method in applications. However, the integer-valued levels are useful with noisy data. The integer coefficients provide solutions, which are easily generalised, but the number of variables in each equation must be fairly small in order to guarantee a successful tuning.

Forcing the model surface to go through the origin may result in unnatural membership definitions. Therefore, a new equation type was developed

$$\sum_{j=1}^{m} A_{ij} X_j + B_i = 0, \qquad (3)$$

The bias term B_i was introduced for fault diagnosis systems.

Several sets of linguistic equations can be combined by matrix presentation

$$A \cdot X + B = 0, \qquad (4)$$

where A is the interaction matrix, B the bias vector and X the vector of linguistic values of the variables. Each row is an individual model, which can be generated separately. Some of the equations may be generated automatically from data, some from the already existing fuzzy set systems, and some are defined directly from prior knowledge.

Using the above defined data, automatic model generation provides a real valued interaction matrix

[-0.632 −0.3435 −1.6525 2.2145].

Effects of different variables can be clearly seen by dividing the matrix by its smallest value giving

[1.84 1 4.81 6.45].

Second variable (conductivity) has the smallest effect on Kappa value. The first variable (refractive index) has 1.8 times bigger effect and the third variable (temperature) 4.8 times bigger effect on Kappa than conductivity. The fourth number in the matrix corresponds to the output (Kappa number). It has the greatest value because all three inputs have an effect on it.

The corresponding model given by the original integer modelling approach is

[-2 1 −6 −6]

The methods are compared using three numerical measures: The correlation between actual measurements and the model, together with the relative error and the squared error of the model. The process data was divided in three parts for training, testing and validation. Training and testing data are 600 and validation data 321 measurements. Collecting the data has encountered several difficulties. Therefore, there is not so much data available and validation data differs slightly from the other sets.

Table 4 shows how the before mentioned linguistic equations behave in modelling. Real valued models seem to lead to somewhat smaller errors than integer valued models. Fig. 2 shows graphical comparison with the validation data.

Table 4. Results from integer valued and real valued Linguistic Equations models.

	Correlation	Relative error	Squared error
Integer valued			
*training	0.75	0.045	1.37
*testing	0.74	0.051	1.56
*validation	0.64	0.039	1.15
Real valued			
*training	0.77	0.033	1.01
*testing	0.70	0.032	0.98
*validation	0.67	0.033	0.96

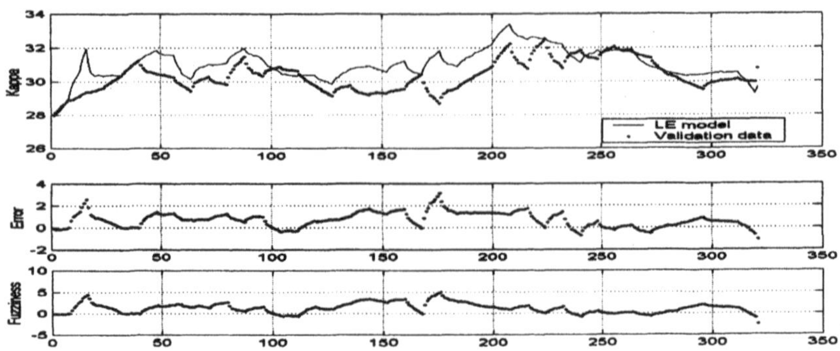

Fig. 2. The comparison between the output of the linguistic equation model and the validation data.

In the next stage the linguistic equations were converted to different kind of fuzzy models. There are several possibilities depending on the following aspects:

- Is the modelling starting from integer or real valued linguistic equations,
- The type of membership functions used in fuzzy models (triangular/trapezoidal)
- The shape of membership functions (linear/non-linear)
- The number of membership functions (linguistic labels) for each variable
- The type of fuzzy rules (Mamdani, Singleton, Takagi-Sugeno).

Following text shows the main results of the comparison. More results are given in (Juuso and Isokangas 2000).

Table 5 shows the results from fuzzy models using triangular membership functions and starting from real valued linguistic equations. Takagi-Sugeno model gives the best correlation and all fuzzy models seem to lead to smaller errors than linguistic equations. This improvement is, however, small compared with the increased complexity of models and the computational load.

Next, the trapezoidal membership functions are used instead of triangular ones. Results are shown in Table 6. Results are, however, worse than with triangular membership functions. The effect is smallest with Takagi-Sugeno model.

Next table shows the effect on the type of linguistic equations. Here integer valued equations are used as a basis for fuzzy models. Results prove to be worse than with real valued equations for Mamdani-type of model. Otherwise they compare well with the results given in Table 5 or are even better.

The tests showed that the number of membership functions or the number of singletons did not have a bigger effect on model performance after some minimum requirements are fulfilled. Neither did the change from linear to non-linear membership functions (Juuso and Isokangas 2000).

Table 5. Results from the fuzzy models. Real valued coefficients, five linear triangular membership functions for each variable leading to 125 rules.

	Correlation	Relative error	Squared error
Mamdani			
*training	0.73	0.037	1.11
*testing	0.57	0.030	0.98
*validation	0.58	0.029	0.84
Singleton			
*training	0.76	0.034	1.03
*testing	0.61	0.035	1.07
*validation	0.63	0.028	0.81
Takagi-Sugeno			
*training	0.78	0.033	1.00
*testing	0.72	0.040	1.17
*validation	0.66	0.029	0.85

Table 6. Results from the fuzzy models. Real valued coefficients, five linear trapezoidal membership functions for each variable leading to 125 rules.

	Correlation	Relative error	Squared error
Mamdani			
*training	0.70	0.038	1.14
*testing	0.55	0.039	1.20
*validation	0.56	0.032	0.96
Singleton			
*training	0.73	0.039	1.19
*testing	0.57	0.050	1.52
*validation	0.56	0.041	1.18
Takagi-Sugeno			
*training	0.77	0.033	1.01
*testing	0.71	0.042	1.30
*validation	0.66	0.032	0.94

4.2
Comparing Linguistic Equations and Neural Network Models

The results gained by linguistic equations were compared to a neural network model that was based on a feedforward network. Levenberg-Marquardt algorithm was used in training. In a model there were two hidden layers and each layer consisted of four neurones. Neural network model reached 0.72 correlation in training and 0.63 in testing. The squared error for model was 0.94 in training and 0.96 in testing. So it compares well with other approaches.

4.3 ANFIS

ANFIS (adaptive neuro-fuzzy inference system) uses input/output data set to construct a fuzzy inference system, whose membership function parameters are adjusted using either a back propagation algorithm alone or in combination with the least squares type of method. This allows fuzzy systems to learn from the data they are modelling. In behaviour, ANFIS is similar to Takagi-Sugeno models.

Table 7. Results from fuzzy models. Integer valued coefficients, five linear triangular membership functions for each variable leading to 125 rules.

	Correlation	Relative error	Square error
Mamdani			
*training	0.75	0.043	1.32
*testing	0.65	0.064	1.93
*validation	0.60	0.042	1.22
Singleton			
*training	0.75	0.037	1.12
*testing	0.64	0.050	1.52
*validation	0.61	0.034	0.98
Takagi-Sugeno			
*training	0.75	0.034	1.04
*testing	0.73	0.035	1.07
*validation	0.60	0.026	0.75

Table 8 shows the results for ANFIS-models with different number of membership functions for each variable. The performance seems to be better than with fuzzy models.

Fig. 3 shows the comparison between the output of ANFIS-model and the validation data. The results are similar to Fig. 2.

5 Conclusions

This paper compares different fuzzy models and neural networks with linguistic equations in modelling the Kappa number in a continuous digester. In training, all methods seem to learn the process behaviour in a similar manner. Differences come out in using the models in process environment. Neural network models are suitable for processes where process conditions are very stable and there is a lot of data available. The Linguistic Equations model works better and it is more suitable for the prediction of the Kappa number, because it is not so sensitive for changes in process conditions. Linguistic equations offer also tools for modelling the system in several operating points and thus developing adaptive models. Since only five parameters are needed for each variable, the Linguistic Equations system can, more easily, be adapted to various operating conditions. This means a big advantage compared with different kind of fuzzy models tested in this study. These lead to a large number of membership functions and rules that are difficult to tune in practice.

Table 8. ANFIS models. There is a different number of membership functions used.

No. of MF/variable		*Correlation*	Relative error	Square error
2 2 2 2	*training	0.81	0.031	0.94
	*testing	0.71	0.036	1.10
	*validation	0.69	0.020	0.59
3 3 3 3	*training	0.82	0.030	0.92
	*testing	0.69	0.035	1.08
	*validation	0.71	0.020	0.59
4 4 4 4	*training	0.84	0.029	0.87
	*testing	0.65	0.044	1.32
	*validation	0.71	0.020	0.59

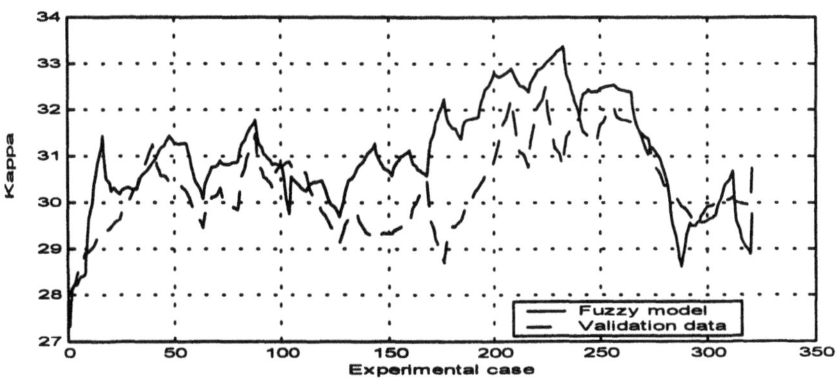

Fig. 3. Comparison between ANFIS-model and the validation data.

The possibilities to use Kappa number prediction in the continuous digester control are different in each control system and process. Analyser measurements can be utilised in the selective alkali dosage and the residual alkali control as well as in the solids profile control when the process makes it possible.

The quality of data is the most important thing in modelling. Data has to be filtered always in some way, and the measurements have to be reliable. Therefore, the analyser or the measurement equipment has to be calibrated and regularly maintained.

References

Haataja, K., Leiviskä, K. and Sutinen, R. (1997): Kappa-number estimation with neural networks. In: Proceedings of IMEKO World Congress, Finnish Society of Automation, Tampere, Finland, Volume XA, pp. 1–5

Juuso, E. K., Leiviskä K. (1995): A Development Environment for Industrial Applications of Fuzzy Control. In: Zimmermann, H.-J. (ed.): Proceedings of the Third European Congress on Intelligent Techniques and Soft Computing –EUFIT'95. Verlag Mainz, Aachen, Volume 2, pp. 796–803

Juuso, E. K. (1996): Linguistic Equations in System Development for Computational Intelligence. In: Zimmermann, H.-J. (ed.): Proceedings of the Fourth European Congress on Intelligent Techniques and Soft Computing –EUFIT'96. Verkag Mainz, Aachen, Volume 2, pp. 1127–1131

Leiviskä, K. (1999): Process Control in Fiber Line. In Leiviskä, K. (ed.): Process Control. Papermaking Science and Technology, Book 14, Fapet Oy, Jyväskylä, Finland, pp. 73–99

Lemmetti, A., Leiviskä, K. and Sutinen R. (1998): Kappa-number Prediction Based on Cooking Liquor Measurement. Report A No 5, May 1998. Control Engineering Laboratory, University of Oulu, Oulu, Finland, 31 p

Murtovaara, S., Juuso, E. K., Sutinen, R., Leiviskä, K. (1998a): Modelling of Pulp Cooking Characteristics by Using Fuzzy Logic and Linguistic Equation Models. In: Proceedings of TOOLMET'98 – Tool Environments and Development Methods for Intelligent Systems. University of Oulu, Oulu, Finland, pp. 90–96

Murtovaara, S., Juuso, E. K., Sutinen, R., Leiviskä, K. (1998b): Neural Networks Modelling of Pulp Digester. In: Dourado, A. (ed.): Proceedings of CONTROLO'98, 3rd Portuguese Conference on Automatic Control. APCA, Portugal, pp. 627–630

Murtovaara, S., Leiviskä, K., Juuso, E., Sutinen, R. (1999): Modelling of Pulp Characteristics in Kraft Cooking. Report A No 9, Control Engineering Laboratory. University of Oulu, Oulu, Finland, 20p

Sutinen, R., Furst, P., Joensuu, I. Leiviskä, K. (1990): Advanced Kamyr Digester Control. In: Proceedings of 24th EUCEPA Conference. SPCI, Stockholm, Sweden, pp. 253–264

Sutinen, R. (1996): Prediction of cooking results using neural networks. Automaatioväylä 5, pp. 19–22 (In Finnish)

Fuzzy Quality Control of a TMP Plant

Jaakko Myllyneva[1], Lennart Karlsson[2], Ismo Joensuu[3]
[1]University of Oulu, Control Engineering Laboratory, Finland, P.O.Box 4300, FIN-90014 Oulun yliopisto, Oulu, Finland
[2]Holmen Paper AB, Hallstavik, Sweden
[3]ABB Industry Oy, Systems Group, Oulu, Finland

1 Introduction

The control of thermomechanical pulping (TMP) refiners has been extensively studied during the past two decades. However, only a small number of TMP refiners have closed-loop quality control (Sui et al. 1998). The basic reason for this situation is the complexity of the refining process. The quality control has to cope with two different kinds of disturbances: slowly proceeding wearing of refiner plates and faster variations in raw material quality (Leiviskä et al. 1993). Information about the pulp quality is usually obtained from laboratory testing of pulp or finished paper samples. Typically, this information is a few hours old and therefore of no use for real-time control. Also, delays in on-line freeness and fibre length measurements have limited the use of automatic on-line quality control.

Generally, refiner control systems are based on the principle that freeness is directly related to the specific energy consumption in the process (Kooi 1994). The plate position controls the refiner motor load based on different strategies given by the refiner manufacturers. The main problem of this principle is the lack of reliable production rate measurement in order to have an accurate main line specific energy calculation. In many cases, the lack of reliable sensors has been tried to compensate for using different modelling techniques, which also have inaccuracies due to assumptions regarding some process variables.

Due to strong process non-linearity, interactions and uncertainties in the quality analysis, conventional control methods turn often difficult to take in use. The modern fuzzy control tools are an attractive way to handle these problems, because they do not require any exact process models, which are difficult to maintain e.g. due to the drifting of the process variables. The rule based fuzzy quality control combines several different operational control strategies of the production personnel. And, when the basic control loops are stabilised on both refining stages there exists a good basis for the actual quality control of the TMP-plant.

The AutoTMP fuzzy quality control system was installed on four refining lines at Holmen Paper, Hallstavik mill. The process includes Sunds Defibrator SD65 refiners on both primary and secondary stages. The installed control system has successfully reduced the variations in freeness and has always maintained the mean fibre length on the acceptable level.

The TMP fuzzy quality control is based on the fact that the refining consistency is a key variable in the TMP-plant control because it affects how energy is transferred to fibres. The TMP Consistency Analyser (TCA) enables an automatic consistency control. The analyser measures the consistency in the blow line directly after the refiner and utilises this information in the control of the dilution water and the refiner motor load. An accurate in-line consistency measurement makes it possible to use the consistency and motor load controls concurrently without causing any slow drifting in process feedback, which might be experienced in the case of model based control systems (Sutinen 1999).

2
Process Description

Traditionally, pulps are divided to mechanical and chemical pulps. Mechanical pulps are produced in grinding and refining processes. In grinding the wood is processed as whole and in refining as chips. Mechanical pulps are mainly used in newspaper together with chemical pulps. Bleached mechanical pulps are also used in different printing papers. Other areas of usage are also packaging, and toilet and wallpapers

The main processes in mechanical pulping are.

- SGW: stone groundwood.
- PGW: pressurised groundwood.
- RMP: refiner mechanical pulping.
- TMP: thermomechanical pulping.
- CMP: chemical mechanical pulping.
- CTMP: chemical thermomechanical pulping.

The growing popularity of the TMP process is due to its high yield and good properties of pulp.

In refining process, the fibres in wood are separated by grinding chips between grooved refiner plates. Chips are fed in the centre of plates from where centrifugal forces convey chips towards the outer diameter of plates. The distance between plates is biggest in the centre and decreases, when moving outwards. So first, chips are crushed to smaller pieces and later separated as individual fibres, fibre bundles, and also to finer fractions. The temperature of wood also increases during refining, which also facilitates the separation of fibres.

The quality of refiner pulp depends on the specific energy consumption during refining and the refining intensity, i.e. energy consumed in plate impacts on fibres. If the specific energy consumption is kept constant, the time delay in refining defines the amount of plate impacts on fibres in the refining zone.

TMP plant number 3 in Holmen Paper, Hallstavik produces 215.000 ton quality pulps annually from Scandinavian spruce. The pulp is produced for paper machine No 11 producing standard and improved News grades. The refining system consists of four main lines and two reject lines. The main lines have two refining stages and the reject lines have one refiner each. All the refiners are SD65 refiners and they are equipped with 10 MW motors. The production capacity is 700 ton/day.

The process flow scheme is illustrated in Fig. 1.

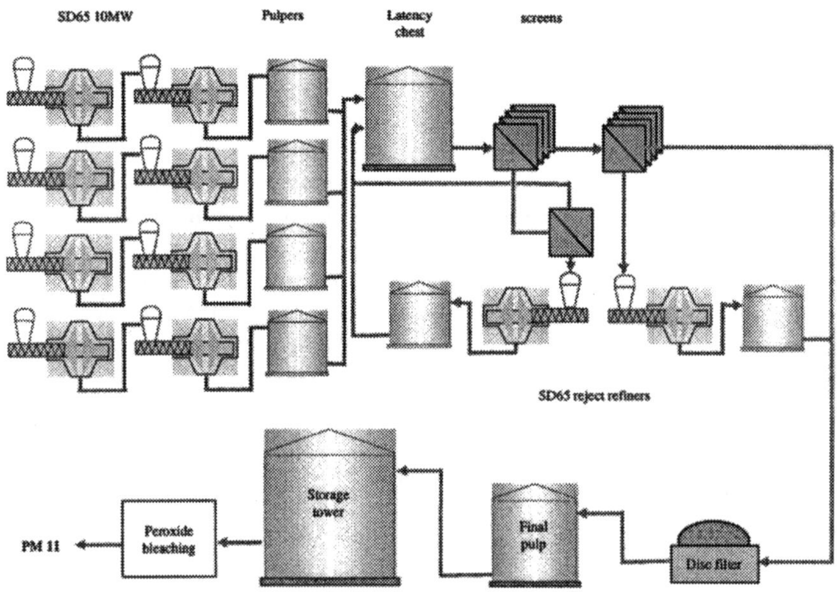

Fig. 1. TMP plant No 3 in Hallstavik.

Chips are screened to remove oversize chips, fines and knots. Accepts are washed to remove sand, stones and other solid materials. The washed chips are sent to the dewatering screens and then fed into the pre-heaters, where lignin is softened to aid defibrillation in refining. Finally, the heated chips are fed through a high-compression plug screw to the primary refiner. The production rate of each refiner line is adjusted by varying the speed of a metering screw. Primary refining is performed at 40–45% consistency. The pulp from the primary refiner passes through a cyclone to separate the steam and then into the secondary refiner. The secondary refiner is operated at the same consistency as the primary stage. The refining pressure is 4.5 bar. Following each secondary refiner, the pulp goes to the pulpers, from where the stock is pumped to a common latency chest. In the latency chest the pulp fibres are prepared for two-stage screening. From screening the rejected fibres go to additional refining in two parallel reject refiners. Accept is dewatered in disc filters and the final pulp is led to a storage tower.

3 AutoTMP Control System

Variations in the motor load are either due to changes in the incoming wood characteristics, varying refining consistency or they are self-generated in the process. In the AutoTMP control strategy the primary refining consistency is first stabilised based on the TCA blow line consistency measurement and consistency control. Next, the motor load control maintains the required motor load constant at the primary stage thereby eliminating the feed changes.

Proper control of the consistency and motor load is important not only for the primary refiners but also for the secondary refiners point of view, since these are typically subjected to the same disturbance patterns as the primary stage. The load and consistency swings in the secondary refiners tend to be even two or three times larger than in the primary stage. This underlines the importance of good control of the primary stage in order to avoid amplified disturbances in the secondary stage. Further stabilisation is achieved through the feedforward consistency control and the motor load control of the secondary stage. The power split control adjusts the set points of the primary and the secondary motor loads (Sutinen 1999).

When the consistencies and motor loads are stabilised on both stages there exists a good basis for the actual quality control of the TMP-plant (Sutinen 1999). The highest level in the AutoTMP control hierarchy is the fuzzy quality control, which maintains freeness and fibre length on the desired levels by adjusting the motor load and consistency targets. The AutoTMP control hierarchy is shown in Fig. 2.

4 Adaptive Consistency Control Strategy

Improved refiner consistency control plays a key role in the TMP operation because consistency affects the viscoelastic properties and wet mass of the wood. Furthermore, consistency affects the centrifugal force in the refiner gap and thus the residence time and further how energy is transferred to fibres.

An automatic consistency control improves the refiner performance and decreases disturbances caused by variations in the chip quality. The TCA-analyser measures the consistency in the blow line directly after the refiner and utilises this information in the control of the dilution water and refiner motor load. The TCA measurement is based on the reflection of infrared light from the pulp stream. Four wavelengths of the IR-spectrum are used to evaluate the pulp consistency. The principal component analysis (PCA) is used in calibration of the TCA signals against the laboratory analysis. The calibration has proven to be very stable and insensitive to fibre size distribution and freeness changes in the pulp stream (Fig. 3).

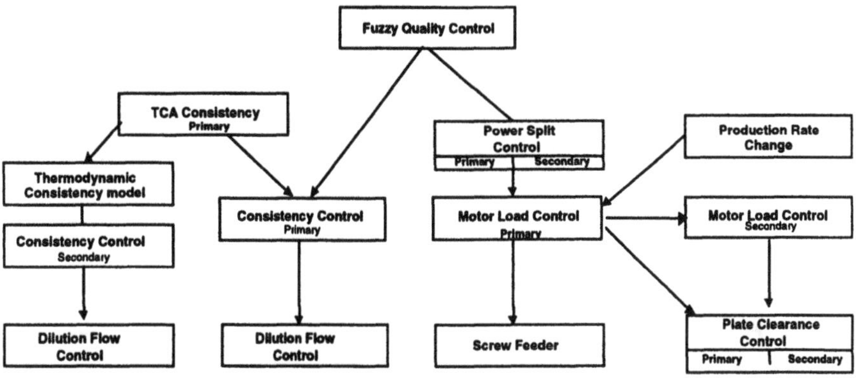

Fig. 2. AutoTMP control hierarchy.

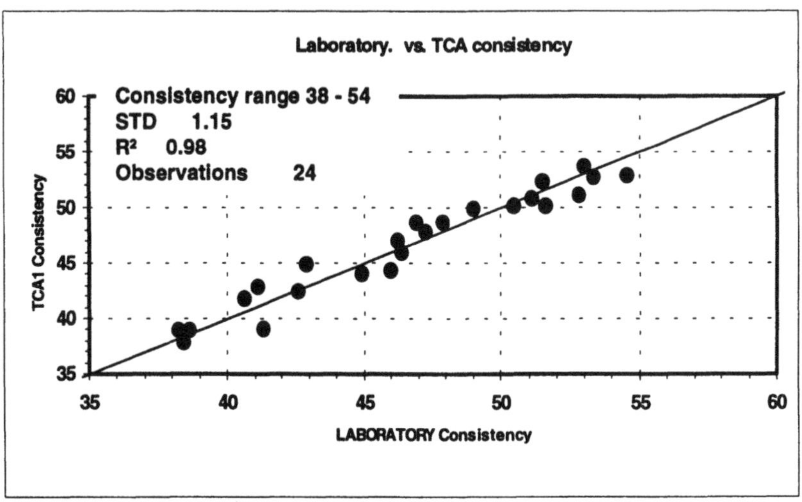

Fig. 3. Correlation between laboratory and TCA consistency in Hallstavik.

The TMP process is typically characterised by time lags between the dilution water (chip feed rate changes) and the reaction of the refiner motor load, and further process responses which change over time due to factors such as changes in the refining gap and segment wear.

In order to run the process close to its limits; the control system has to react quickly enough to disturbances. Because of changes in the process dynamics and static process gains, conventional PID controllers turn out to be unsuccessful at fulfilling this requirement. An adaptive controller was therefore selected as the basic element of the AutoTMP system. Another advantage of this system is that it can be commissioned in a considerably shorter time. The adaptive controller adjusts itself continuously to changes in the process dynamics and also includes the built-in dead-time compensation and continuous adaptive feedforward control. The block diagram of the adaptive consistency control is shown in Fig. 4 (Sutinen and Joensuu 1996).

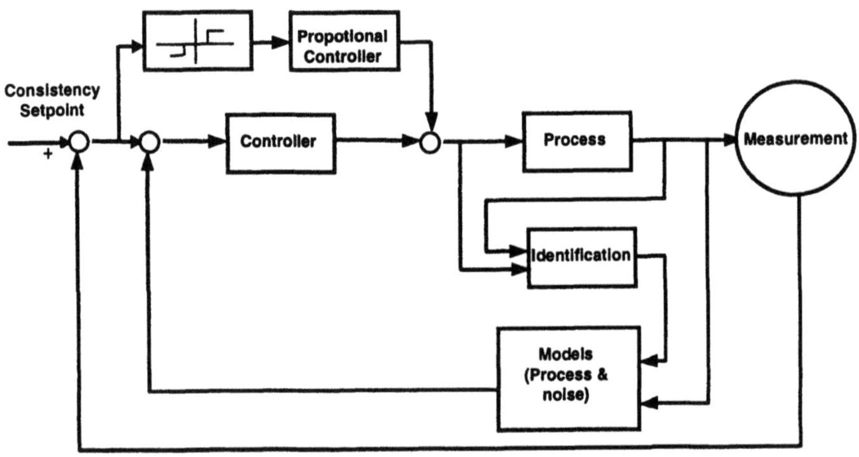

Fig. 4. Block diagram of the adaptive controller (Sutinen and Joensuu 1996).

The secondary stage consistency is calculated based on the primary stage consistency and heat and material balances.

5 Motor Load Control

The refiner control system is not the simplest one to put into practice, since the process is non-linear and its behaviour varies according to the refining gap, segment wear, and production rate, etc. Results of identification studies have shown that a change in these variables leads to significant changes in the process gain.

When the consistency is first stabilised adjusting the plate gap or the chip screw speed or both can control the motor load. The motor load control (Fig. 5) maintains the constant primary refiner load by changing the set point of the rotation speed of the chip feeder, the refiner plate position or both depending on the operation practice in use. The chip feeder speed and plate position are independent controls that respond individually to changes in the refiner motor load. The motor load control is the master control for these two, and it co-ordinates the actions of the actuators so that the response to load changes is appropriate. The operation does not depend on whether just one or both of the actuators is included in the motor load control. Both of these actuators have separate tuning factors that determine how fast and when the control will respond. A normal characteristic of refiners is that the incremental gain between the motor load and plate gap is subject to a slow drift due to plate wear and even sudden changes in its sign due to pulp pad collapse. Such a pad collapse can be caused by a change in the operation point, or may occur suddenly due to a feed rate or consistency disturbance. In such a case when the plate gap decreases, the motor load increases until a "pinch-off" point occurs. When the pulp pad collapses, the motor load drops significantly. In order to avoid such a situation the motor load control uses other measurements to secure a safe operation of refiner plates. These measurements include plate acceleration, plate vibration, plate touch point and plate position.

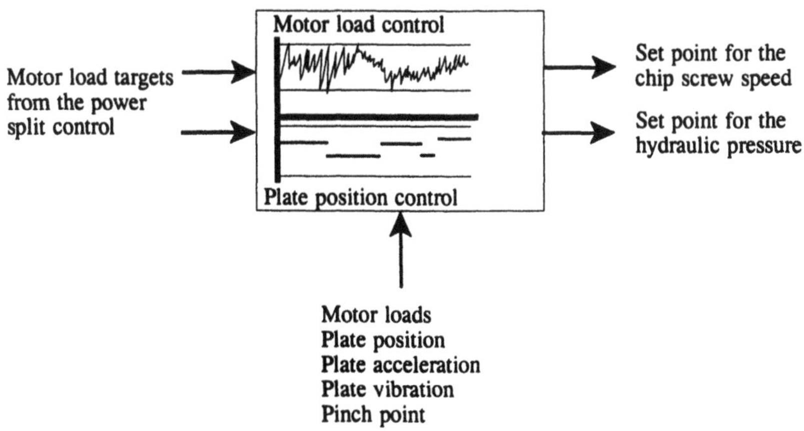

Fig. 5. Motor load control.

6
Fuzzy Quality Control

The fuzzy quality control is based on on-line measurements of freeness, fibre length and TCA consistency (Fig. 6). The measurements in Hallstavik are carried out with a PQM analyser, and the measurement cycle per one sampling point is about 15 min. Sometimes the uncertainty of quality measurements can even be larger than the real variations in the pulp. Therefore statistical analysis of the quality measurements is needed before they are used in the control.

The fuzzy quality controller (Fig. 6) works as a master controller adjusting the set points of the motor loads and the primary stage consistency. Inputs to the fuzzy quality controller are the difference between the freeness set point and measurement, freeness set point, consistency set point and mean fibre length measurement. The consistency set point is used as a control input instead of the measurement, because the actual consistency always includes short-term fluctuations around the set point and therefore may cause some unnecessary output changes. The fuzzy quality controller needs to know the consistency level to be able to decide if there is still room to adjust the consistency set point within its limits.

The fibre length measurement is used as the highest priority input variable to the quality controller. If the fibre length decreases below the given limits the controller corrects the freeness set point in order to keep the fibre length on the acceptable level. Freeness has a target value, and it is controlled by the motor loads in the both stages and the primary consistency.

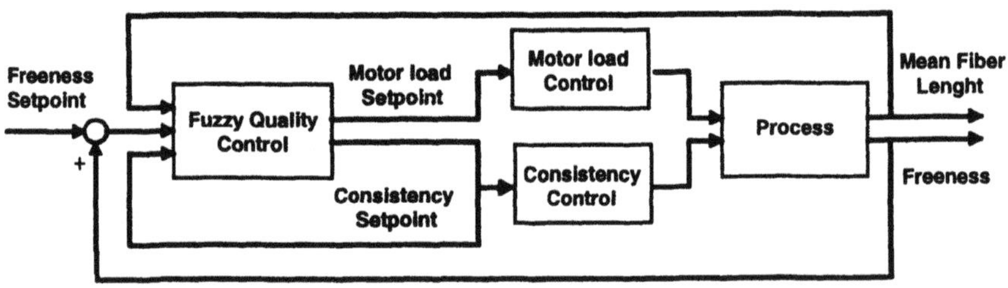

Fig. 6. Fuzzy quality controller.

If the fuzzy controller observes that the fibre length is becoming shorter it first tries to correct the situation by increasing the consistency. If the corrections are not enough the controller corrects the freeness set point. A temporary increase of the freeness set point allows the controller to decrease the motor load set points thus helping the system to return the fibre length above the minimum length allowed. When disturbances are over, the freeness set point is moved back to the original target.

The rules used in the fuzzy quality control are based on the operating strategy shown in Fig. 7. The generated rules are based on extensive mill trials and process knowledge. Fig. 7 shows different situations, how freeness and mean fibre length can fluctuate and also strategies, how to reach the target values. For example, in a situation where freeness is too high and fibre length is ok, the freeness needs to be decreased without shortening the fibres. By increasing the consistency, relatively more fibres are in the plate gap and therefore refining energy requirement increases. It will increase the motor load, which typically is compensated for by opening the plate gap. The consequence can be that the freeness level will not decrease as planned. A better strategy is that the plate gap is kept constant by increasing the motor load set point at the same time with the consistency set point. In this way freeness can be decreased.

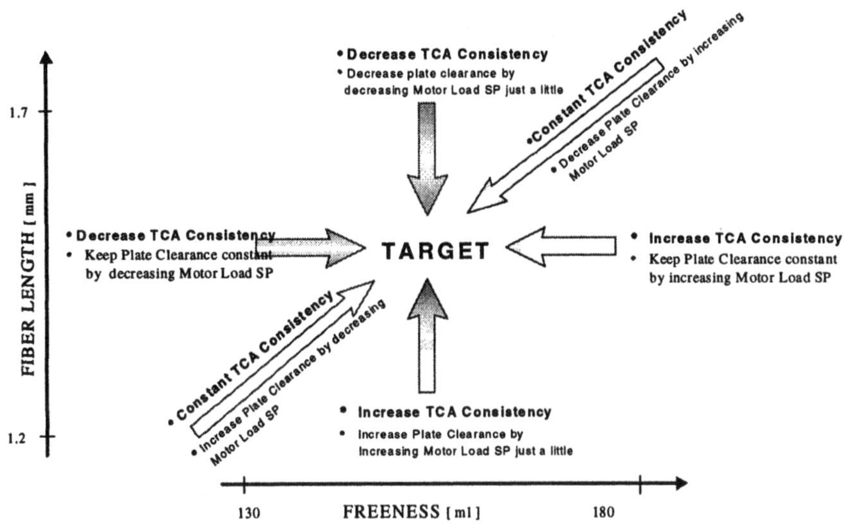

Fig. 7. The basic operation strategy of the fuzzy quality controls.

The fuzzy quality control system in Hallstavik includes 19 rules to maintain the stable quality. Freeness has three linguistic values. The freeness measurement can be above, below or at its set point. In this case, the fibre length has only two linguistic values, ok and short, because in practice it can never be too long. Consistency has three labels.

The rules used in the example situation described earlier depend on the actual consistency level. If the consistency is OK the rule used is:

If Freeness is HIGH and Consistency is OK and Fibre Length is OK then INCREASE A LOT Mload-set-point and INCREASE A LOT Consistency-set point.

If the consistency is LOW the rule used is:

If Freeness is HIGH and Consistency is LOW and Fibre Length is OK then INCREASE A LOT Mload-set-point and INCREASE A LOT Consistency-set-point.

If the consistency is already high, and it is not permitted to increase it more, then the rule used is:

If Freeness is HIGH and Consistency is HIGH and Fibre Length is OK then DON'T CHANGE Mload-set-point and DECREASE A LOT Consistency-set-point.

Decreasing consistency may mean that fibres are shortened, but consistency levels are selected in a way that in a normal situation fibre length stays always above the desired level. Further, in an abnormal situation the freeness set point is increased to prevent fibres getting too short. Usually, two or more rules are acting at the same time and the executed control action is a combination of them.

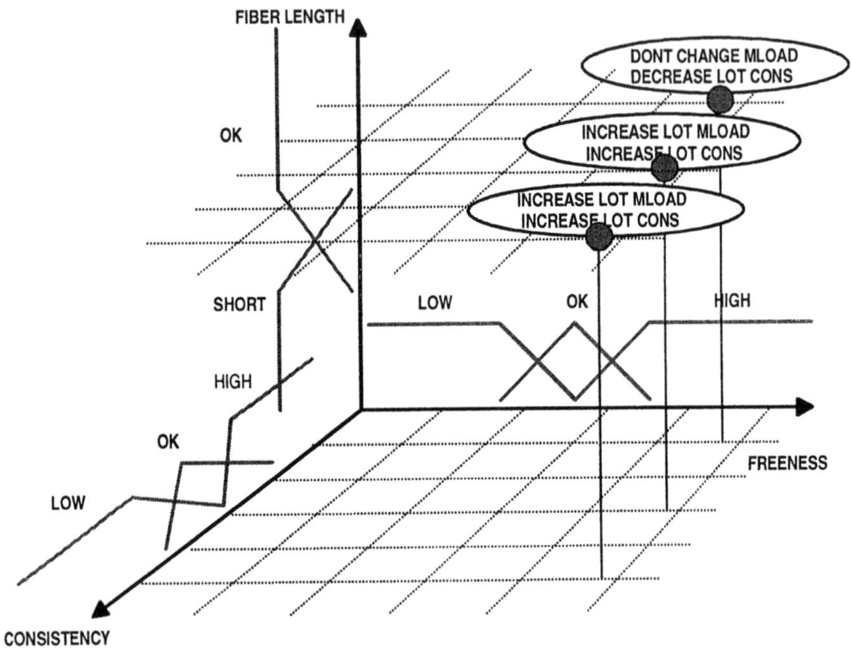

Fig. 8. Example of the fuzzy control rules.

The marked dots in Fig. 8 present three different rules. Freeness is always HIGH and fibre length OK. When the consistency is HIGH control actions are different than when the consistency is LOW or OK.

Due to non-continuous quality analysis available the fuzzy controller is event based using a variable control period. Freeness and mean fibre length are measured with a PQM in-line analyser. The analyser's measuring cycle is about 10 minutes and control actions are done after three samples. The newest sample has the highest weight when the samples are averaged. The interval between the control operations is about 30 minutes due to the process dynamics and delay caused by the analyser sampling frequency. In practise, the control period can vary from

20 minutes to 2 hours, depending on how many other sampling points the PQM analyser is monitoring.

The fuzzy quality controller has been implemented in the mill's automation system using standard fuzzy controller toolbox available in the system. Minimum-maximum method was used in inference and singletons were used for outputs

Any fuzzy controller has to be tuned mostly based on experience. Because the quality control is slow in this particular process, it was necessary to test the rules and tuning parameters before the implementation. Using process data collected earlier and knowledge of the process a test simulator was configured (Myllyneva et al. 1999). The simulator model describes on a general level, how the specific energy and consistency affect the freeness and fibre length. Also, basic values for tuning parameters were studied, which decreased the time required for the tuning.

7 Results

The installed fuzzy quality control system includes a consistency analyser after the primary refiner on four main lines. Adjusting the dilution water controls consistency. The motor load is then controlled by adjusting the primary refiner plate gap and the chip feeder speed. Next, the fuzzy quality controller corrects the consistency and motor load set points based on the freeness and mean fibre length readings and the developed fuzzy logic rules.

A six months follow-up period demonstrated 95–100 % uptime of the motor load and consistency controls and over 90 % uptime of the fuzzy quality controls, respectively. Furthermore, the fuzzy quality controls have improved the stability of the process compared to the basic refiner control operations. As an example, results from several evaluation periods demonstrated a decrease in standard deviation of the process variables listed in Table 1 when the fuzzy quality controls were switched on.

Table 1. Process stabilisation in line 5

	Decrease in std. dev.
Freeness	30 – 50 %
Consistency	20 –30 %
Motor Load	20 –30 %

Fig. 9 shows an example how the fuzzy quality control affects the line 5 freeness during the evaluation period.

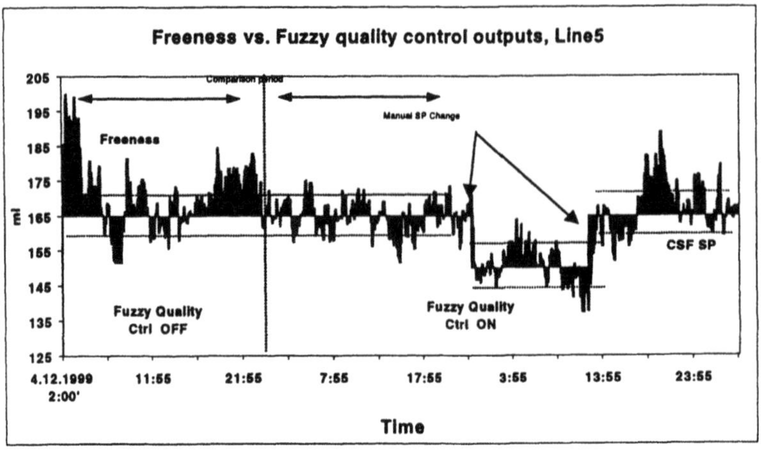

Fig. 9. The effect of fuzzy quality control on the freeness on line 5.

Generally, the fibre length was on a good level over the whole evaluation period. Fig. 10 shows a situation where the measured mean fibre length has dropped suddenly. There are two main reasons for this type of fast fibre shortening. Firstly, it might have been caused by chip quality changes, or there might have been fast drifting in the fibre length measurement due to dirty water. As can be seen, the fuzzy quality control has responded correctly using the rule: "freeness is ok and fibre length is short" (Fig. 11). The control has gradually decreased the motor load set points and simultaneously increased the consistency set point until the fibre length has raised back to the acceptable level.

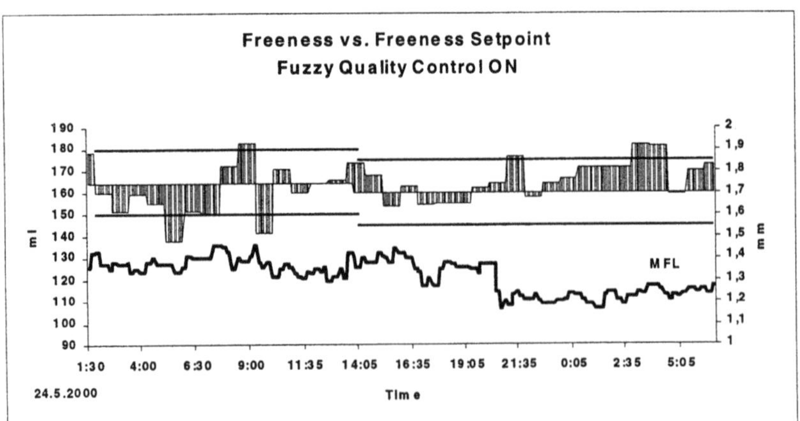

Fig. 10. Freeness (upper curve) and mean fibre length (MFL).

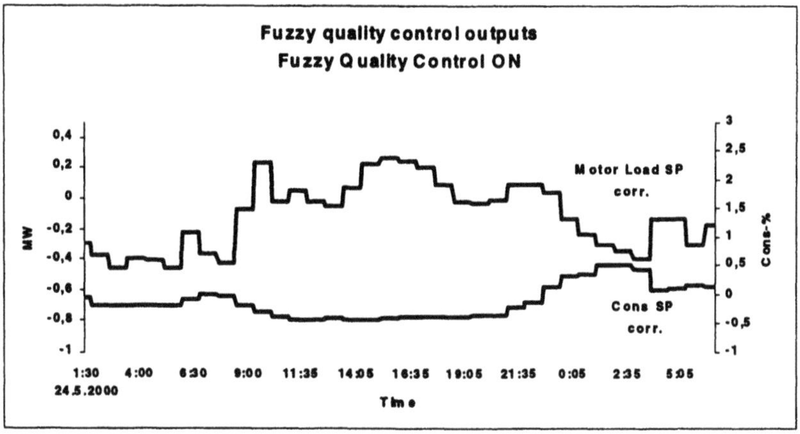

Fig. 11. Behaviour of control variables in the example of Fig, 10. Note that the decrease in motor load and increase in consistency after the time 21:00 are caused by the sudden decrease in mean fibre length shown in Fig. 10.

The fuzzy rules have been tuned so that the main control variable is the fibre length. Therefore, the freeness level is allowed to increase, when the motor load control opens the plate position more than the consistency controller requests. When the fibre length was corrected back to its normal value, the other rules will have a bigger influence on the controller's output, and therefore the fuzzy quality controller has started to increase the motor load gradually in order to get the freeness close to its set point.

8 Discussion

In the thermomechanical pulping process the ability to measure the final pulp quality is still limited to secondary phenomena's of its characters. It limits, in its part, the ultimate goal of any quality control to produce uniform sheet properties with significantly reduced energy consumption levels. However, with today's technology it is possible to stabilise the refining consistency and freeness, maintain the fibre length and optimise the energy consumption, thus improving the sheet properties, too.

The fuzzy quality control makes the operating practices uniform and levels freeness variations between different shifts. Further, the quality control responds quickly every time to changes in the freeness and eliminates the production of off-spec pulp.

Actual energy savings were difficult to verify in this period due to varying process conditions, but it is clear that when the freeness is stabilised energy peaks caused by freeness fluctuations are eliminated.
A future goal to improve the fuzzy quality control is to tune the fuzzy rules to meet even better the energy saving demand. The key question for improved rules is to find out the best compromise between the fibre length and the energy efficient consistency level for the current process operation. The optimum consistency level will not remain the same all the time because of the changing process conditions and disturbances.

9 Conclusions

The AutoTMP control system including fuzzy quality control modules has been successfully applied to four refiner lines at Holmen Paper, Hallstavik mill. The control system features the TCA consistency measurement in the blow line. Based on it the refining consistency and motor load are controlled using adaptive control methods. Finally, 19 rules have been used to describe the fuzzy control strategy where the fibre length is always maintained on an acceptable level and the freeness is controlled to be on a target level.

The fuzzy quality control requires on-line freeness and fibre length measurements, TCA consistency measurement and consistency and motor load controls in the first stage. Secondary stage consistency and motor load controls are recommended as further enhancements.

Operator confidence in the control system has been very high; 95–100 % uptime for the consistency and motor load controls and over 90% uptime for the fuzzy quality controls have been experienced. The six months follow-up period proved a 30–50 % decrease in the freeness standard deviation. Next, when the fuzzy quality control has stabilised the freeness level on the target, more attention can be paid to further develop rules of energy efficient refining.

References

Kooi, S. B. L. (1994): Adaptive inferential control of wood chip refiner. Tappi Journal 77(11):185–189

Leiviskä, K., (ed.) (1999): Process Control. Papermaking Science and Technology, Book 14. Fapet Oy, Jyväskylä, Finland

Leiviskä, K., Sutinen, R., Saarinen, K. (1993): Optimal Quality Control TMP-plants. In: 18th International Mechanical Pulping Conference, Proceedings. EUCEPA, Oslo, pp. 245–251

Myllyneva, J., Kortelainen, J., Sutinen, R., Leiviskä K. (1999): Fuzzy Quality Control of TMP Refiner Line. In: 6th International Conference on New Available Technologies. SPCI. Stockholm. pp. 280–283

Sui 0. S., Sanche, L., Mills, C., Smith, W., Smith, G., Douglas, T. (1998): Model based pulp quality control of TMP refiner. In: Pulping Conference Proceedings. Tappi Press, Atlanta, FL Book 1, pp 295–308

Sutinen, R.A. (1999): Optimized TMP-plant control. In: Advanced Technology from Finland for the Forest Industry. Finnpro, Montreal

Sutinen, R.A., Joensuu, I. (1996): Experiences of TMP refiner control based on an in-line consistency measurement. In: 50th Appita Annual General Conference. Appita, Carlton, Victoria, Australia

Intelligent Supervisory Control of an Industrial Rotary Kiln

Esko Juuso[1], Mika Järvensivu[2], Olli Ahava[3]
[1]Control Engineering Laboratory, University of Oulu, P.O. BOX 4300, FIN-90014, University of Oulu, Finland
[2]Pronyx Control Software, Veneentekijäntie 4, FIN-00210, Helsinki, Finland
[3]UPM-Kymmene, P.O.BOX 42, FIN-68601, Pietarsaari, Finland

1 Introduction

The process industries face considerable control challenges, especially in terms of the consistent production of high quality products, more efficient use of energy and raw materials, and stable operation under different conditions. Flexibility and fast reactions to market situations and changing operating conditions are necessary. Interactions between control loops make multivariable systems non-linear. The important quality variables can be estimated only from other measured variables. The physical limitations of actuators must be taken into account. The different time delays depend greatly on operating conditions and can dramatically limit performance and even destabilise the closed loop system. Uncertainty is an unavoidable part of process control in real world applications. These demands cannot be met by traditional control techniques only, and several methodologies have therefore been developed to extend the applicability of control systems (Juuso 1999a).

The complex dynamics and multi-variable nature of the calcination process, with its non-linear reaction kinetics, long time delays and variable lime mud feed characteristics, make the lime kiln process inherently difficult to operate efficiently. During its operation many interconnected variables must be considered and control actions must be designed to meet multiple and sometimes conflicting objects, and changing operating conditions. Furthermore, some of the measurements are unreliable, and the kiln characteristics may change during a long run. The lime kiln operation may also be upset by disturbances such as changes in the composition and/or properties of the lime mud. In addition, certain process variables must be maintained within predefined constraints in order to assure safe operation of the kiln process.

Supervisory control of the kiln process is thus in many respects a demanding task. Lime kilns are therefore frequently one of the last areas in a pulp mill to be automated and most of the kilns have been and are still operated without a supervisory control system. The absence of closed loop controls results in inefficiencies in energy consumption and variations in lime quality. Furthermore, manual control of the process, increase the probability of equipment failures and increase the risk of flue gas emissions. In contrast, there are outstanding improvement potentials associated with the improved control. Firstly, energy efficiency can be improved considerably, and the benefits associated with a more stable lime quality can also be significant when kiln control is improved (Bailey and Willison 1986; Crowther et al. 1987). The costs associated with repairing the refractory lining can also be reduced if the number of high temperature excursions is reduced, as reported in Dekkiche (1991) and in McIlwain (1992). Emissions can also be decreased when the occurrence of process upsets is eliminated, or at least reduced, by means of improved control (Uronen and Leiviskä 1989; Järvensivu 1999).

To sum up, the state of the art in the rotary kiln control is that the systems based on the fuzzy logic and rule-based approach have already proven their applicability for controlling industrial kilns and systems combining various intelligent control and prediction techniques will be the future trend. In this paper, the structure and main functions of the supervisory-level kiln control system, which has been implemented at the Wisaforest pulp mill in western Finland, is presented. In the modular and hierarchically structured system, non-linear models are used in conjunction with advanced high-level controllers based on a novel linguistic equation approach. In the present system, earlier neural network models described in have been replaced with linguistic equation models. Finally, the results obtained during the five-month testing period of the system are discussed and an assessment of the achieved benefits presented.

2 Process Description

The causticising process, which is a part of chemical recovery at the pulp mill, produces white liquor from green liquor. This process consumes lime, CaO, and produces lime mud, principally $CaCO_3$, as a by-product. The purpose of calcination is to convert the mud back into lime for reuse in the causticising. The prime method used for calcination in the pulp industry is a rotary kiln. Conceptually, the lime kiln, which is a large cylindrical oven, can be divided into four process zones according to the temperature profile of the solids material and the flue gas in the kiln. These four zones are drying of the wet mud, heating of the mud powder up to the reaction temperature, calcining of calcium carbonate into calcium oxide, and agglomeration of the formed fine powder into granules. The endothermic calcining reaction (Eq. 1) starts at about 800 °C, and the reaction heat required is 1.8 MJ/kg $CaCO_3$ decomposed.

$$CaCO_3\ (s) + heat \rightarrow CaO\ (s) + CO_2\ (g) \tag{1}$$

At the Wisaforest mill, the lime mud is first washed and then dewatered by means of two parallel vacuum drum filters. After the filters, the mud at about 75%

dry solids is conveyed to a screw feeder, which distributes it into the flue gas duct. The flue gas then carries the mud into a cyclone, which is called the lime mud drier (*LMD*). Simultaneously, heat is transferred from the flue gas to the wet mud, which evaporates off the residual moisture. The dried and preheated mud powder is discharged from the bottom cone of the cyclone and fed into the cold-end of the kiln. After the *LMD* cyclone, the flue gas passes through an electrical precipitator and two-stage venture type wet scrubber. The dust, which is captured in the electrical precipitator, is fed back into the kiln together with the preheated mud. The mud powder then moves down the gradient of the kiln as a result of the inclination and rotation, and the burnt lime is discharged from the kiln through the product coolers located at the hot-end of the kiln.

About 80 % of the total heat energy input into the kiln is obtained by gasifying sawmill dust in a circulating fluidised bed gasifier and then burning the generated gas in a burner installed at the hot-end of the kiln. Heavy fuel oil is only used as a secondary fuel. Primary combustion air, which is taken between the stationary hood and the product coolers, is forced into the kiln together with the fuel. Secondary combustion air is induced into the kiln by a low negative pressure maintained by an induced draft fan located after the electrical precipitator in the kiln exhaust system. The resulting counter-current flow of the hot combustion gases from the hot-end to the cold-end of the kiln distributes the heat energy along the length of the kiln.

3 Rotary Kiln Control

The control of rotary kilns has been studied since the early 70s with both phenomenological and empirical methods. Expertise and fuzzy logic were introduced to these applications already in very early stage. In addition to fuzzy logic and rule-based systems, rotary kiln applications based on the model predictive control approach have been reported in the literature during the 90s.

3.1 Phenomenological Models

The first applications were models based on the dynamics of the solids phase, the fundamental principles of heat transfer mechanisms, and the kinetics of drying, heating and calcining reactions. The developed models were used to estimate the temperature of solids, flue gas and refractory along the length of the kiln. These models, and many of the models since developed, have provided a useful insight into the rotary kiln process and also increased our understanding of the interactions and time delays inherent in the rotary kiln process (Uronen et al. 1975; Brown and Rastogi 1983; Smith 1991).

Although phenomenological models have been relatively successful in simulating the operation of a rotary kiln very few, if any, of these models have been extended from the simulation stage to the control of an industrial scale kiln. The main reason for that have been difficulties to achieve an adequately accurate model of the rotary kiln process.

3.2 Empirical Models

Data-driven modelling approaches are based on general function approximators ("black-box" structures), which should capture correctly the dynamics and non-linearity of the system. The identification procedure, which consists of estimating the parameters of the model, is quite straightforward and easy if appropriate process data is available. The structure and parameters of these models do not necessarily have any physical or chemical significance, and therefore, these models cannot be adapted to different operating conditions.

For lime kiln applications, the main emphasis turned already on 70s to developing supervisory control systems based on empirical models, as described e.g. in Uronen and Aurasmaa (1979). The first commercial supervisory systems for the lime kiln was developed on the basis of these studies and its first industrial applications appeared already at the end of the 1970s in two Finnish pulp mills (Elsilä et al. 1979; Sutinen 1981).

Other rotary kiln control systems based on the empirical models were also reported during the 1980s in the Unites States (Bailey and Willison 1985).

Computational intelligence provides various techniques for data-driven modelling. A neural network based system has also been tested for the control of the simulated kiln and also proposed for an industrial application, as reported in Ribeiro and Correia (1995).

A linguistic equations approach (LE) has been introduced to dynamic simulators developed for controller design (Juuso 1999b). The fuzzy LE approach combines various information sources including mechanistic models, data-driven learning algorithms and expert knowledge to dynamic approximators, state space models and semi-mechanistic models.

Understanding of the underlying phenomena and structure of the model are dealt with interactions, and non-linearity of the system is described by membership definitions tuned by input-output data. Interactions, membership definitions and effective time delays depend on working point conditions. This makes it possible to model changing production conditions. Delays are handled as smoothly changing fuzzy numbers. In the multi-model approach, the smooth switching between various sub-models provides a robust platform for testing controller alternatives. The robust dynamic simulator is continuously used in developing adaptive industrial controllers.

3.3 Expertise in Rotary Kiln Control

Human operators can handle many processes that are very difficult to handle for conventional control methods. Rule-based programming is commonly used in the development of expert systems, but this paradigm leads to serious maintenance and testing problems in practical applications where rule-based systems become really massive.

Fuzzy logic control (FLC) has been widely used in non-linear and multivariable control (Juuso 1999a). Fuzzy sets provide a unified framework for taking into account the gradual or flexible nature of variables and the representation of incomplete information. For small fuzzy logic controllers, modelling the operator's actions is a quite useful procedure. For complex systems, acquiring the knowledge from the human operator is a tedious and time-consuming task and other methods must be considered.

Some experiments with automatic control based on deterministic decision tables were carried out already before fuzzy control. For a cement plant, these decision tables were inspired by instructions found in a textbook for kiln operators, which contained the control rules for manual operation. First experiments with real cement plants were done since 1972 in Denmark (Oestergaard 1993). FLS Automation introduced fuzzy logic to the control of a real cement kiln in 1978. The rule base was modified and the first two industrial applications, the Danish cement plant and a rotary kiln in a Swedish paper mill, were implemented at the end of 1979 (Ostergaard 1993). Other industrial, fuzzy logic based kiln control applications have since been reported (Sheuer 1995; Nilsson 1997).

The first rule-based expert system for kiln control was developed in 1982 and, since then, the system has been further developed as described in Haspel (1991) and in Dekkiche (1991). Other rule-based systems for controlling rotary kilns have also been developed and reported (Hall 1993 and Hagemoen 1993).

The kiln process has been extensively studied at the Wisaforest mill. A kiln control system based on fuzzy logic was developed during 1993. Further details on the system, which was formulated on the basis of operator interviews and implemented in the *DCS*, can be found in Penttinen (1994) and in Ruotsalainen (1994).

Encouraged by the promising results, the research was continued with the main emphasis on applying a novel linguistic equation approach for fuzzy modelling and simulation of the kiln process (Juuso et al. 1996) and to extend the control to the whole kiln process.

The original knowledge-based fuzzy system has been modified in order to take into account high level control and adaptation to changing operating conditions. Various models have been used in the tuning of the controllers (Juuso 1999b). A need for continuous adaptation to changing operating conditions has been clearly seen.

3.4
Model-Based Control

Model-based control is widely used in industrial applications but a lot of identification work with local linear models is needed. Multiple model adaptive control (MMAC) allows different control structures, i.e. each mode corresponds to one model and one controller. Switching control strategies are based on selecting a controller from a finite set of fixed controllers.

In model-based predictive control (MPC), prediction of future plant behaviour is used to compute the appropriate control action at each sampling instant over a prediction horizon. Various kinds of models can be used in MPC. Simulation results in rotary kiln applications based on the model predictive control approach have been described in Charos et al. (1991) and in Zanovello and Budman (1999), while the results of an industrial application can be found in Smith and Aggarwal (1998) and also in Valiquette (1999).

3.5
Adaptation to Operating Conditions

Gain scheduling is an intuitively appealing approach, which provides a gradual adaptation of the gains but is limited to a fixed control structure. Typically, a slow variation of system dynamics is assumed, which may lead to difficulties in compensating for fast disturbances. Sliding mode controllers (SMC) are designed to provide convergence of a system's trajectory to a sliding surface, which is based on plant dynamics. Drastic control changes leading to chattering are the main obstacles to the use of SMC controllers in the process industry. Adaptation algorithms suitable to these techniques are not easy to construct for lime kiln control.

Classical adaptive schemes do not cope easily with pronounced and fast changes unless the adaptation rate is made very high. One approach in these cases is to use a switching control scheme, which selects a controller from a finite set of predefined fixed controllers. Each controller can have a completely different structure. The switching strategy can be based on heuristic rules or predictions with models. A wide variety of models and controllers needed in rotary kiln applications would make the switching complicated.

3.6
Hybrid Approach

Intelligent methods have extended the toolbox of hybrid, semi-mechanistic or gray-box modelling. Non-linear systems can be approximated with reasonably accurate fuzzy models, and for small systems, models can be inverted into controllers. For large systems, a fuzzy model can predict the process behaviour, and can be used in model-based predictive control (MPC).

A rule-based kiln control system, in which neural networks are used to represent the rule set, has been reported in Yang et al. (1997). Recently, a control system incorporating a predictive-adaptive controller to maintain the kiln process close to the set points being generated by an expert system, was described in Barreto (1997). In the same paper, neural networks in combination with the expert system were proposed as a potential candidate for future development.

The tradition of FLCs has been continued with linguistic equations. The linguistic equation approach was first used as a development and tuning methodology for fuzzy set systems (Juuso et al. 1996). The first direct industrial application was implemented in 1996 (Juuso et al. 1998), and recently, the number of applications has increased rapidly in modelling, intelligent analysers, control and fault diagnosis (Juuso 1999a).

4
Control Environment

4.1
Field Survey

A field survey of the operation of the kiln process with special attention paid to the reduced sulphur emissions, was carried out concurrently at the mill. The results of the survey showed, in addition to a considerable enhancement potential in the performance of the kiln process, that improved control is also a feasible means of reducing kiln emissions (Järvensivu et al. 1999a). The research was then continued with the aim of developing, on the basis of the accumulated process knowledge, a supervisory-level control system for the kiln process. Preliminary results of the kiln control system are reported in (Järvensivu et al. 1999b; Sievola 1999; Järvensivu and Jämsä-Jounela 1999c).

4.2
Overall Structure of the Kiln Control System

The lime kiln control system at the Wisaforest mill consists of basic instrumentation and on-line analysers, a process automation system, an information system and an intelligent supervisory control system. The basic instrumentation of the kiln process includes for instance electromagnetic flow meters, radioactive density meters, pressure sensors, thermocouples and a pyrometer. In addition, on-line analysers are used for measuring excess oxygen and sulphur dioxide of the flue gas and total reduced sulphur (*TRS*) emissions.

The automation system (*ALCONT*), which is a *DCS* supplied by Honeywell-Measurex, provides an interface to the basic instrumentation and the flue gas analysers, and also carries out the handling of alarms, safety interlocks and logic sequences, as well as basic control loops. The information system, *XIS* supplied by

Neles Automation, is used to collect all the kiln process measurements from the automation system, and it provides also access to the results of the laboratory analysis. The information system also performs the calculations of the average values at different time intervals, and records both the current and historical data into the real time database. The major measurements, the frequently performed laboratory analysis, and the most important basic control loops of the kiln process are listed in Tables 1 and 2, respectively.

In the kiln process at the Wisaforest mill, as in all the other lime kiln processes as well, the primary output variables such as the residual $CaCO_3$ content and causticising efficiency of the burnt lime, are not measured on-line. For instance, at the Wisaforest case the burnt lime is manually sampled with an 8-hour time interval and the residual $CaCO_3$ content of the sample is analysed with a delay time of about 2 hours in the laboratory. The laboratory analysis is significantly delayed and therefore impracticable from the control point of view. Therefore, measurable state variables with known correlation with the primary process outputs need to be chosen as controlled variables.

The excess oxygen content of the flue gas after the LMD cyclone, the temperature of the flue gas in the cold-end of the kiln and the hot-end temperature measured by means of a thermometer were selected as controlled variables. The excess oxygen and cold-end temperature are closely related with the flue gas heat losses and emissions. The hot-end temperature affects first and foremost the burnt lime quality, but also the heat energy consumption. The draft fan speed and sawdust feed rate were selected as the manipulated variables of the stabilisation-level of the system due to a strong and predictable correlation with the controlled variables. The rotational kiln speed was also selected as a manipulated variable, because it provides a mean for the control of the residence time of the solids.

Table 1. Major measurements and laboratory analysis

On-line measurements	Laboratory analysis
- Lime mud flow rate, l/s	Lime mud feed;
- Lime mud density, kg/dm^3	- Dry solids content, %
- LMD temperature, °C	- Residual alkaline, %
- Cold-end temperature, °C	- Residual CaO, %
- Hot-end temperature, °C[1)]	
- Burning zone temp., °C[2)]	Burnt lime;
- Cold-end pressure, mbar	- Residual $CaCO_3$, %
- Kiln drive torque, %	- Causticising efficiency, %
- Excess oxygen , %	
- TRS emissions, ppm	

[1)] Thermometer, [2)] Pyrometer

Table 2. Basic-level control loops (in DCS)

Controlled variables	Manipulated variables
- Lime mud flow rate, l/s	- Feed pump speed, rpm
- Lime mud density, kg/dm^3	- Dilution valve position, %
- Draft fan speed, rpm	- Inverter (fan motor), rpm
- Kiln rotational speed, rpm	- Inverter (kiln drive), rpm
- Sawmill dust feed rate, %	- Inverter (feeder speed), %
- Fuel oil flow rate, kg/s	- Valve position, %

5 Stabilising Controls

The supervisory control system is implemented using a Windows NT workstation and G2, which is an object-oriented environment for developing and deploying intelligent systems, from Gensym. The supervisory control is based on the *StabOpti* and *LECont* concepts described later in this paper. Its main variables are shown in Table 3.

The lowest level of the system provides bi-directional interface to the low-level systems through a bridge to the *RTDB* of the information system. The lowest level also calculates certain values required at a higher level in the application. The next level in the hierarchy, i.e. stabilisation-level, has been designed to ensure stable operation of the kiln process over the entire production rate range. In practice, it is done by means of appropriate changes made in the set points of the basic-level control loops, which are implemented in the DCS system.

The purpose of the highest level, i.e. optimisation-level, in the hierarchy is to optimise the kiln process operation in respect to the burnt lime quality, kiln energy efficiency and environmental protection by means of determining optimum target values for the stabilising-level controllers.

5.1 *StabOpti* Concept

StabOpti is an intelligent supervisory-level control concept for stabilising and optimising continuous industrial processes (see Fig. 1 for a scheme of the process stabilisation and optimisation). The system is divided into inter-related and hierarchically structured modules.

The lowest level of the hierarchically structured system provides an interface to the external systems. It also carries out raw input data validation and application-specific calculations required at a higher level in the system.

Table 3. Controlled and major constraint variables of the supervisory level system

Controlled variables	Manipulated variables
- Excess oxygen , %	- LMD temperature, °C
- Cold-end temperature, °C	- Cold-end pressure, mbar
- Hot-end temperature, °C	- TRS emissions, ppm

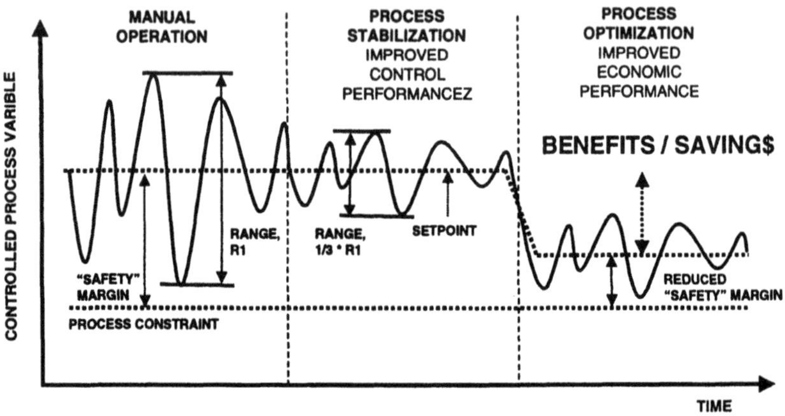

Fig. 1. Schematic presentation of the process stabilization and optimization.

The next level in the hierarchy, i.e. stabilisation, has been designed to ensure stable operation of the process over the entire production range. It comprises modules for the feedforward compensation of load disturbances based on the predetermined non-linear models obtained from process data, and for the rejection of unmeasured disturbances by means of the stabilising controllers based on a novel *LECont* concept.

The stabilisation level also provides functions for protecting personnel, equipment and the environment when preset constraints are exceeded, e.g. in the case of severe disturbances and/or abnormal process conditions. The handling of constraint problems is based on structured natural-language rules and procedural reasoning.

The purpose of the highest level, i.e. optimisation, in the hierarchy is to optimise the kiln process operation with respect to product quality, energy efficiency, environmental protection, etc. It is done inferring the state of the process and then determining optimum target values for the stabilising controllers. This is also the level dealing with economic benefits of the system. Optimisation is dealt with in Chapter 6.

5.2 *LECont* Concept

The *LECont* concept is based on linguistic equations in which traditional fuzzy systems described by if-then rules are represented by matrix equations and non-linearity are handled by membership definitions (Juuso et al. 1996). The first direct linguistic equation controller was implemented in 1996 for a solar power plant in Spain (Juuso et al. 1998).

The basic level of the *LECont* concept is based on a PI LE controller represented by a single linguistic equation

$$\Delta u = e + \Delta e \,, \tag{1}$$

which is a special case of the matrix equation AX = 0 with the interaction matrix A = [1 1 −1], and variables $X = \begin{bmatrix} e & \Delta e & \Delta u \end{bmatrix}^T$. A real-valued LE controller provides a very smoothly changing control surface (Juuso, 1999a). Feedforward controllers can be constructed by inverting steady state models. Non-linearity is introduced into the system by membership definitions, which corresponds to membership functions used in fuzzy logic controllers. Operation of the controller is modified by variables describing operating conditions, and the working point is defined by a working point equation.

For large disturbances, a braking action was implemented by means of a correction, which tries to lower the speed of error compensation even before the error gets to zero. The braking action is automatically activated when the error becomes large enough, and the system returns to normal operation when the fast trend is removed. The changes in operating conditions can modify membership definitions considerably. Since a different asymmetry is needed in different operating conditions, an asymmetrical correction defined by the trend variable is included. If several trends affect the offset, an asymmetry equation can be introduced. This action is restricted to fairly stable conditions and is not used simultaneously with the braking action.

The *LECont* concept has been implemented in G2. The compact system corresponds to a three-level cascaded controller:

- The basic PI type LE controller handles the normal operation with symmetrical membership definitions.
- The operation condition controller changes the control surface of the basic LE controller by modifying the membership definitions for the change of control variable Δu.
- The predictive LE controller changes the membership definitions for the derivative of the error Δe. This level contains both the braking and asymmetrical actions.

The strength of the braking and asymmetrical actions can be modified if the operating conditions are clearly changing. Process delays should be taken into account in selecting the strength of braking. Membership definitions are required for error, change of error, change of control, original error, trend, correction coefficient and all the working point variables. Modularity is beneficial for the tuning of the controller to various operating conditions, and most important is that the same controller can operate over the whole working area. The LE approach has also been extended to dynamic modelling and simulation (Juuso 1999b).

5.3 Modules of the Stabilisation-Level

The stabilisation level comprises the modules of the feedforward control models, stabilising controllers and constraint handling. The main purpose of the feedforward models (FFMs) module is to take care of production rate changes. The FFMs module manages production rate changes by means of appropriate adjustments made in the set points of the basic-level control loops (SPs) in the kiln process. The purpose of the stabilising controllers (SCs) module is to provide adaptation in case of inaccuracy in the models and after unmeasureable disturbances in the kiln process inputs or the process itself by means of small corrections in the SPs. On top of the SCs module, the constraints handling (CHs) module is activated for protecting personnel, equipment and the environment when pre-set constraints are exceeded, e.g. in the case of severe disturbances and/or abnormal process conditions. In addition, the CHs module is used to tackle large deviations from the target conditions by means of stepwise changes in the SPs.

The *FFMs* module relies on the predetermined relationships obtained from the large amount of process data collected at the Wisaforest. While, the SCs and CHs modules rely on real-time inference of the state of the kiln process with respect to the target values (that are determined at a highest level in the system) as well as preset constraints.

5.3.1 FF Models Applied During Production Rate Changes

The model used for the feedforward (FF) control of the kiln rotational speed complies with the relationship between the production rate and kiln rotational speed obtained from the kiln manufacturer. In line with the recommendations, the rotational speed is increased along with increasing production rate in order to maintain a constant filling rate in the kiln. In addition, the LE model takes into account cold-end pressure, which is used indirectly to indicate undesired ring formation in the cold end of the kiln and the kiln load predicted by means of the torque of the kiln drive.

The purpose of the LE models used for the FF control of fuel supply and draft fan speed is to provide sufficient heat energy into the kiln and ensure a proper flow of combustion air into the kiln when the production rate is raised, and vice versa.

The inputs of the LE model used for the FF control of the draft fan speed are the 150-minute moving average of the production rate and the 180-minute moving average of the cold-end pressure. The draft fan speed is increased to ensure a proper flow of the combustion air into the kiln, i.e. an appropriate level of the excess oxygen in the flue gas, when the production rate is raised, and vice versa. The other input in the model, the cold-end pressure, is used to compensate the restrictive influence of the existing ring on the flue gas flow, and hence also on heat transfer from the burning zone to the cold-end of the kiln.

Two separate LE models are applied for FF control of the sawdust feed rate depending on the rate of heavy fuel oil burning. The first model is used when sawdust is burned alone, and the other model when heavy fuel oil is burned as a secondary fuel in addition to sawdust. In both cases the sawdust feed rate is altered in accordance with the 150-minute moving average of the production rate. The first model also takes into account the 150-minute moving average of the torque (load) of the kiln drive, which is an indirect indication of the amount of solids in the kiln. This is done in order to provide sufficient heat energy to the kiln for heating the solids up to the reaction temperature, and for calcining the calcium carbonate into calcium oxide, while ensuring that the lime is not overburned. In the other model, the 10-minute moving average of the flow rate of fuel oil is used as a second input in order to maintain the total heat energy input into the kiln despite the changes made to the fuel oil flow rate.

5.3.2 Stabilising Controllers and Constraint Handling

On top of the FF models, small corrections to the set points of the fuel supply and draft fan speed are made by means of the stabilising-level controllers on the basis of the excess oxygen content of the flue gas and the temperature profile along the length of the kiln. Furthermore, if the LMD temperature is below the pre-set limit then the draft fan speed is increased in order to eliminate the risk of increased emissions. The draft fan speed is also increased if the excess oxygen level is already low while, at the same time, the fuel supply into the kiln is increased either by the system or by the operator. In the case of a large deviation in the hot-end temperature, or fast changes in the quality of the sawdust used as primary fuel, stepwise correction to the sawdust feed rate is made in order to avoid extensive temperature excursions.

The draft fan speed is corrected at 10-minute scan intervals on the basis of the excess oxygen content of the flue gas and the temperature profile along the length of the kiln, by means of a LE controller. The LE is applied because an appropriate correction to the draft fan speed is often a compromise between conflicting corrections required to bring the excess oxygen and the cold- and hot-end temperatures closer to their target values.

An increase in the draft fan speed is used to increase the excess oxygen, raise the cold-end temperature and lower the hot-end temperature, and vice versa. In addition, structured natural language rules are used to check the LMD temperature each time a new measurement is obtained. If it is below the pre-set limit, then CH module increases the draft fan speed in order to eliminate the risk of increased emissions. In this way also the formation of acid condensates in the flue gas channel is avoided. Both phenomena are due to the too low temperature during lime mud drying. The CH module is also used to increase the draft fan speed if the excess oxygen level is already low while, at the same time, the fuel supply into the kiln is increased either by the system or by the operator.

The sawdust feed rate is corrected at 5-minute scan intervals in order to maintain the cold- and especially hot-end temperatures at the target level. The small corrections to the *SP* of the sawdust feed rate are also defined by a *LE* controller. While, changes in the operating conditions are taken into account by means of the working point *(WP)*. The *WP* is used to scale the corrections according to the operating conditions of the kiln process. For instance, the corrections of the sawdust feed rate is at their largest when the production rate is high, and vice versa, because at a high production rate larger control actions are required to correct the same deviation as at a low production rate. In addition to the *WP*, the consumption of burning air in the sawdust gasifier, which correlates with the sawdust quality *(SDQ)*, is used for scaling the corrections. Smaller corrections are made with high-quality sawdust than with low-quality sawdust. Scaling the determined corrections on the basis of the *WP* and *SDQ*, provides a certain level of adaptivity and hence enables the single *LE* controller to be tuned for different operating conditions. In the case of a large deviation in the hot-end temperature, or fast changes in *SDQ*, the *CH* module is however activated for making stepwise correction to the sawdust feed rate in order to avoid extensive temperature excursions.

6
Optimisation of the Kiln Process Operation

The improved stability of the process obtained by the stabilisation level of the system allows the kiln process to be operated closer to the constraints, and therefore also closer to the optimum. In order to realise the benefits of the reduced variability, the target values of the controlled variables need to be shifted closer to the constraints. Control of the kiln process within a narrowed safety margin, in the presence of unmeasurable changes in the process, necessitates additional supervision and corresponding high-level adjustments to the target values for fulfilling the objectives set on lime quality and energy efficiency as well as the limit set on emissions. Those adjustments required maintaining the process close to the optimal state is computed Those adjustments required to maintain the process close to the optimal state is computed on the basis of the actual kiln operating conditions, e.g. the production rate and moisture content of the mud fed into the kiln.

The target value for the hot-end temperature is determined by means of the NN model, which was developed on the basis of the process data. The model takes into account the 150-minute moving average of the production rate, and the load of the kiln drive as well as the 10-minute average value of the heavy fuel oil flow rate. The target value is increased in order to maintain the burnt lime quality within the optimum range when the production rate or the torque of the kiln drive increases, and vice versa. The burning of heavy fuel oil affects the shape and length of the flame, and the proportion of cold secondary burning air also increases. As a result, the burning of heavy fuel oil has an influence on the range of the hot-end temperature readings, and hence the target also needs to be adjusted.

The target value for the cold-end temperature is adjusted on the basis of the 150-minute, moving average of the production rate, the LMD temperature, which correlates with the moisture content of the lime mud, and the cold-end pressure by means of an NN model. The model is used for calculating the adjustments needed to maintain the heat content of the flue gas at a proper level for drying and heating up the wet lime mud fed into the kiln. The target value is however lowered if the LMD temperature is high enough, and increased accordingly when the LMD temperature decreases below the pre-set limit.

The target value for the excess oxygen is adjusted on the basis of the 120-minute, moving average of the TRS emission, which is not influenced by the rate of air infiltration before the measurement point. If the TRS emissions are low enough, the target value is lowered with small steps in order to keep the heat losses caused by the hot flue gases leaving the kiln at the lowest possible level. In the opposite case, the target value is increased in order to avoid exceeding the limit if the emissions start to rise.

7 Results

The results presented in this chapter are collected during three test periods:

- Data collected between 12/96–2/98 representing manual operation
- Data collected between 2/99–6/99 representing the pilot system
- Data collected between 8/00–10/00 representing the system described in this chapter

7.1 Average Runtime of the System

The supervisory system was connected on-line at Wisaforest mill for the first time at the end of February 1998. The system, which was described in this paper, has been in operation since the beginning of May 2000. The average runtime of the system between March 1998 and October 2000 is shown in Fig. 2 on a monthly average basis. The average runtime in the closed loop mode was about 45%

between March 1998 and January 1999 during the incremental development of the system. The average runtime in the closed loop mode has been over 80% between February 1999 and June 1999 during the five-month testing of the system. Correspondingly, the average runtime in the closed loop mode has been about 95% between August 2000 and October 2000 during the three-month testing of the system.

The average runtime is calculated as the mean time when the control loops of the flue gas fan speed, the sawdust feed rate and the rotational speed of the kiln have been in the closed loop mode divided by the time that the kiln process has been in operation (lime mud federate into the kiln > 15 t/h) and the sawdust has been burned alone or together with heavy fuel oil.

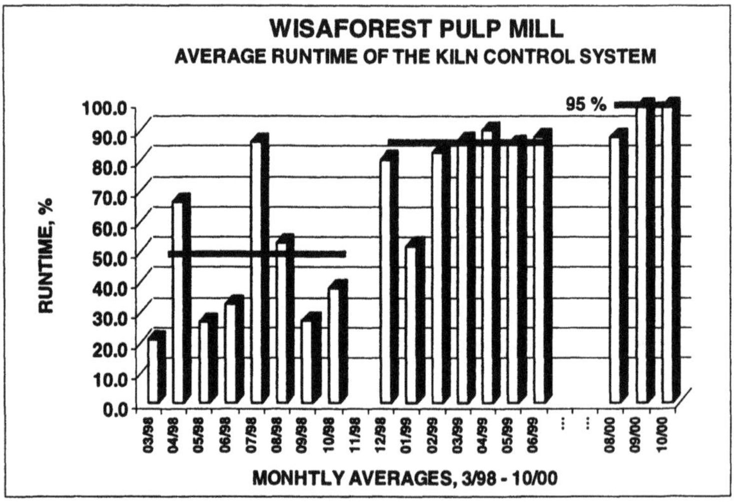

Fig. 2. Average runtime of the intelligent control system of the lime kiln between march 1998 and October 2000

7.2
Dynamic Performance of the System

The dynamic performance of the system was verified using the data collected during the continuous four-day run in the closed loop mode at the beginning of April 1999. Fig. 3 presents the production rate (lower chart), and the hot-end temperature and excess oxygen (upper chart) as one-hour averages. A 24-hour period (marked with the dashed line in Fig. 3) was selected for further analysis.

The period includes at first a production rate change, and then an extended period of kiln operation at constant production rate. Both controlled variables have been kept relatively close to their target values despite the substantial change in production rate during the period concerned, which demonstrates the ability of the system to track the target values of the controlled variables even in the face of considerable load disturbances.

The 24-hour period also includes a fast fall in temperature caused either by a blockage in the sawdust feeding system or by deterioration in sawdust quality. Nevertheless, the system responded to the fall in temperature with an immediate and relatively large increase in the sawdust feed rate and a gradual decrease in the draft fan speed.

As a result, the system was able to prevent any further fall in temperature and subsequently to bring the temperature back into the appropriate range. The fast response, and especially the success in levelling out the temperature variations after the excursion, demonstrates the system's capabilities for effective disturbance rejection.

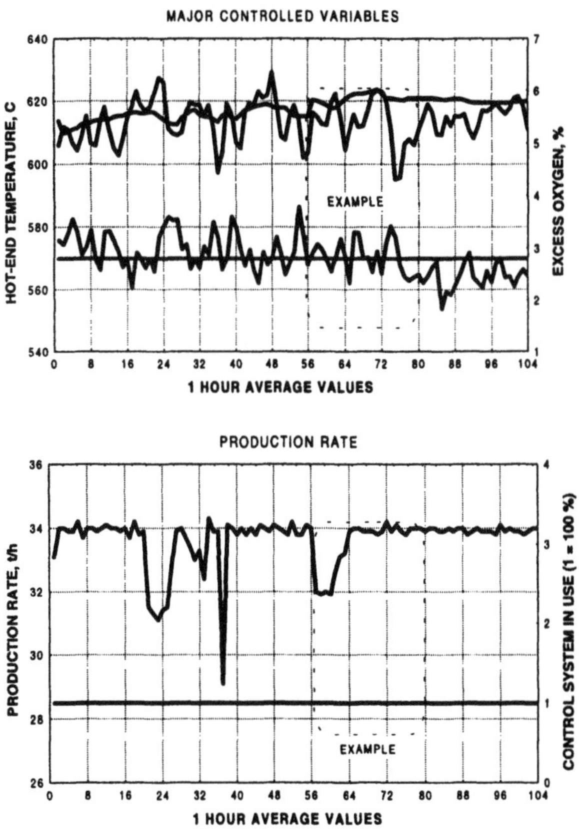

Fig. 3. The continuous four-day runtime span in the closed loop mode. The dashed area of 24 hours has been selected for more detailed study.

7.3 Operational Results and Assessment of Benefits

The operating results were evaluated by means of statistical analysis of the data collected during the manual operation and of the corresponding data obtained during the five-month testing period of the system. According to the statistics, the mean value of the excess oxygen has been reduced by about 15% and the quartile range has been reduced by more than 20%. The mean value of the hot-end temperature measured by means of a thermometer has been reduced by nearly 40°C and the quartile range and the standard deviation have declined by nearly 50% and more than 30%, respectively (see Fig. 4).

The benefits obtained were assessed using data collected during manual operation (between December 1996 and February 1998) and the corresponding data obtained during the first five-month testing period of the pilot system (between February and June 1999) and the second testing period (between August and October 2000). According to the assessment, the main quantifiable benefit during the first testing periods of the system in an economic sense was a notable reduction in heat energy consumption. During the first testing period, the mean of the specific heat energy consumption (5.5 GJ/t_{CaO}) was nearly 7% lower than the respective value during manual operation (see Fig. 5).

During the second testing period of the system the major concern was the performance of the production maximisation module. According to the performed analysis, the mean of the production rate of the kiln process, was 32.9 tons per hour (t/h) in manual operation, 33.2 t/h during the first test period and 33.9 t/h during the second test period. The overall mean of the production rate has been incresed over 5 %, as also illustrated in Fig. 6.

In addition to the easily quantifiable energy savings and increase in production rate, improvements in lime quality have also been obtained. This can actually be assumed on the basis of the significant reduction shown in the variability of the hot-end temperature (the quartile range and the standard deviation of the hot-end temperature have declined nearly 50 % and more than 30 %, respectively). From the ecological point of the view, the main quantifiable benefits were a decrease of over 10% in total reduced sulphur (TRS) emissions and a reduction of about 50% in the proportion of high emission periods (Järvensivu et al. 2000).

8 Discussion and Conclusions

Adapting to changing operating conditions is necessary in the process industry, as clearly demonstrated by the test results. A simple fuzzy controller easily outperforms PID controllers, and the operating area can be extended considerably by higher level structures: the high-level control provides tools for selecting control objectives, and the switching control expands the selection of the control strategies. In principle, the predefined adaptation could be implemented by cascaded FLCs, but these are very time-consuming to tune and also to maintain (Juuso 1999a).

A basic linguistic equation controller corresponds to a normal FLC, and the working point control extends the operation to adaptive fuzzy control. The braking and asymmetrical actions correspond to predictive switching control. All these properties are included in the same controller, which operates smoothly since the adaptation of membership definitions, the switching between special modes and the objective selection are performed simultaneously and gradually. Because of process delays, complicated self-tuning algorithms are not included in the controller implementation. Understanding the meaning of different operating conditions is essential for successful tuning.

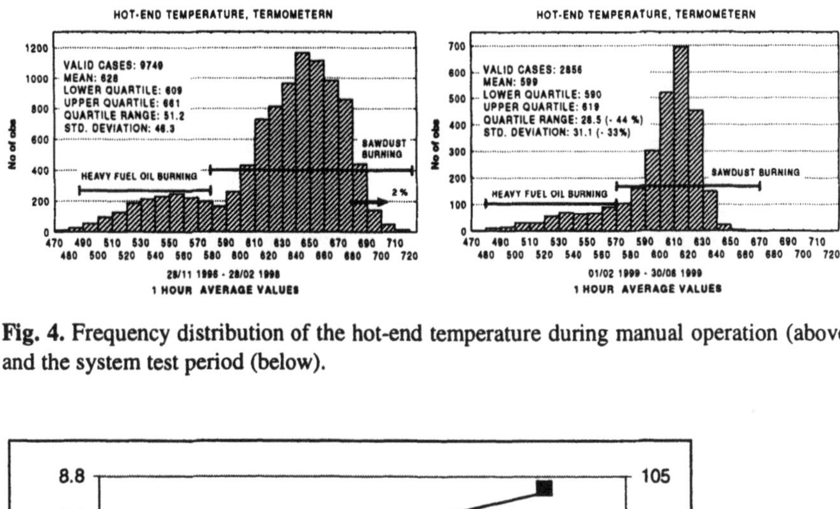

Fig. 4. Frequency distribution of the hot-end temperature during manual operation (above) and the system test period (below).

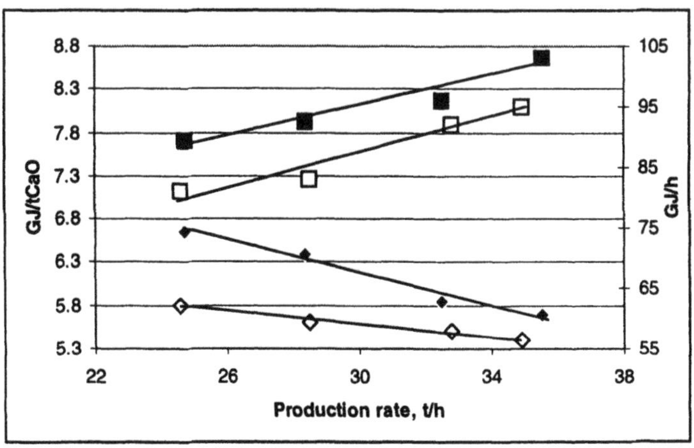

Fig. 5. Heat energy consumption as a function of production rate during manual operation (black symbols) and the test period (white symbols). Upper curves are for energy consumption in GJ/h.

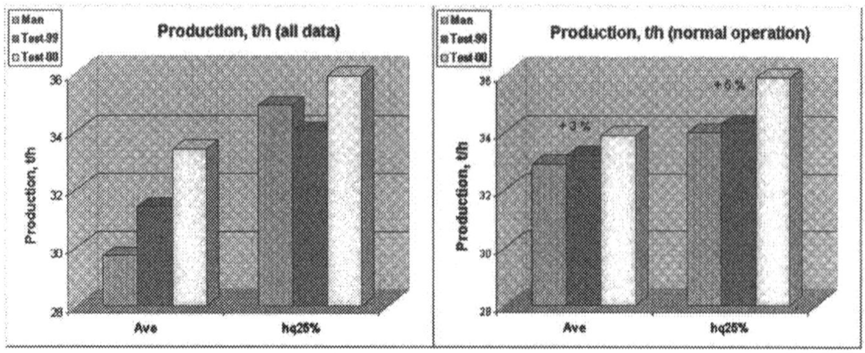

Fig. 6. The development of the production rate during three test periods. The uppermost picture includes all data and the lowermost data collected during the normal operation. The denotation "hq25%" includes periods with the high production.

The directions and strengths of models and control actions can be understood on the basis of process expertise. Linguistic equation systems can also be translated into rules and fuzzy set systems. The *LECont* concept also has a close link to advanced conventional control systems. Working point models control the strength of actions, which could be considered as gain scheduling algorithms using a specific LE structure. Smooth switching with fuzzy methods is an important extension to switching control. The concept provides a compact implementation for extending the properties of control systems. The *LECont* concept can also be extended to model-based control since the LE approach can be applied to dynamic modelling and simulation.

In the *LECont* concept, rejection of the unmeasured disturbances is adapted to operating conditions in a robust way. Improved dynamic operation of the process provides a basis for determining the targets for the stabilising control. The opportunities to reduce the safety margins have been utilised at the optimisation level by the *StabOpti* concept. Smooth changes of target values provided by the *StabOpti* concept are also beneficial for the operation of the *LECont* concept at the stabilising level.

The ecological and economic benefits obtained by the combined use of the new concepts *StabOpti* and *LECont* are considerable. The survey has also shown an improvement in operational flexibility. Furthermore, the operators' workload has fallen and they have gained a better insight into their process. Of more general relevance, the experience gained during the project has shown that experimental knowledge of process behaviour, together with a sophisticated and complete IT solution, can be applied to improve process efficiency and produce considerable benefits.

References

Bailey, R.M., Willison, T.R. (1986): Supervisory control for lime kilns slashes operating costs by up to 20 %. Pulp Paper 60(2), 100–105

Barreto, A.G. (1997): Lime kiln hybrid control system. In: Proc. IEEE Workshop on Dynamic Modeling Control Applications for Industry. NewYork, USA, pp. 44–50

Brown, K., Rastogi, L. (1983): Mathematical modeling and computer control of lime kilns. In: Tappi Pulping Conference Proceedings. Tappi Press, Atlanta, pp.585–592

Charos, G.N., Taylor, R.A., Arkun, Y. (1991): Model predictive control of an industrial lime kiln. Tappi Journal 74(2), 203–211

Crowther, C., Blevins, T., Burns, D. (1987): A lime kiln control strategy to maximize efficiency and energy management. Appita Journal 40(1), pp. 29–32

Dekkiche, E.A. (1991): Advanced kiln control system. Zement-Kalk-Gips 44(6), 286–290

Elsilä M., Leiviskä K., Nettamo K., Pulkkinen T. (1979): Computer control of causticization and lime kiln area is possible, Pulp Paper 53(12), 152–155, 159

Hall, M.B. (1993): Kiln stabilization and control - a COMDALE/C expert system approach. In: 35th IEEE Cement Industry Technical Conference Proceedings, Toronto, pp. 201 - 218

Haspel, D.W., Taylor, R., Brooks, J. (1991): A new perspective on the cement making process through LINKman. In: Proc. Cemtech Conference. Praque, Czechoslovakia, pp. 1.3.1–12

S. Hagemoen, A. (1993): An expert system application for lime kiln automation, IEEE 6 pp. 91–97

Juuso, E.K. (1999a): Fuzzy Control in Process Industry: The Linguistic Equation Approach. In: Verbruggen, H.B, Zimmermann, H.-J., Babuska, R. (eds.) Fuzzy Algorithms for Control, International Series in Intelligent Technologies. Kluwer, Boston, pp. 243–300

Juuso, E. K. (1999b): Intelligent Dynamic Simulation of a Lime Kiln with Linguistic Equations. In: ESM'99: Modelling and Simulation: A Tool for the Next Millennium, 13th European Simulation Multiconference Proceedings. SCS, Delft, The Netherlands, 1999. Volume 2, pp. 395–400

Juuso, E., Ahola, T., Leiviskä, K. (1996): Fuzzy logic in lime kiln control. Proceedings of TOOLMET '96 Symposium – Tool Environments and Development Methods for Intelligent Systems. University of Oulu, Oulu, Finland, 111–119

Juuso, E. K., Balsa, P., Valenzuela, L. (1998): Multilevel Linguistic Equation Controller Applied to a 1 MW Solar Power Plant. In: Proceedings of the 1998 American Control Conference – ACC'98, Philadelphia, PA, June 24–26, 1998, volume 6, pp. 3891–3895

Järvensivu, M., Kivivasara, J. Saari, K. (1999a): A field survey of TRS emissions from a lime kiln. Pulp Pap. Canada 99(11), 111–115

Järvensivu, M., Jämsä-Jounela, S.L. (1999c): Intelligent control system for the lime kiln. Proc. 14th IFAC Word Conference, Beijing, China, 361–366

Järvensivu M. and Seaworth B. (1998): Neural network model used for quality prediction and control. In: IFAC 1998 International Symposium on Artificial Intelligence in Real-Time Control (AIRTC'98). IFAC, Arizona, USA

Järvensivu, M., Juuso, E., & Ahava, O. (2000): Intelligent supervisory-level control of industrial processes. Paperi ja Puu 82(6), 386–391

McIlwain, J.A. (1992): Kiln control. Pulp and Paper Canada 93(11), 34–37
Nilson, L. (1997): Skoghall brings cement logic to causticizing. Pulp Pap. Eur. 2(10), 24–27
Penttinen, R. (1994): Fuzzy control of the lime kiln. Thesis for Master of Science (in Finnish). Control Engineering Laboratory, University of Oulu
Ruotsalainen, J. (1994): Control of the pulp mill chemical recovery circuit. Thesis for Licentiate in Technology. Control Engineering Laboratory, University of Oulu
Ostergaard, J.-J. (1993): Fuzzy control of cement kilns a retrospective summary. In: Zimmermann, H.-J. (ed.): Proceedings of EUFIT'93 – First European Congress on Intelligent Techniques and Soft Computing. ELITE Foundation, Aachen, p. 552–562
Ribeiro, B.M., Correia, A.D. (1995): Lime kiln simulation and control by neural networks. In: Neural Networks for Chemical Engineers. Elsevier Science Ltd., Amsterdam, 163–191
Scheuer, A., & Principato, M. (1995): Experience with the PYROEXPERT kiln control system at the Leimen cement, ZKG International 48(9), 464–471
Sievola, H. (1999): Hot end temperature control of a lime kiln based on linguistic equations. Thesis for Master of Science (in Finnish). Control Engineering Laboratory, University of Oulu
Smith, D.B. & Aggarwal, P. (1998): Advanced lime kiln control: Operating results from a mill installation. In: Proc. Tappi Process Control, Electrical & Information Conference). Tappi Press, Atlanta, USA, pp. 347–353
Smith, D.B., Edwards, L. (1991): Dynamic mathematical model of a rotary lime kiln. In: TAPPI Engineering Conference Proceedings. Tappi Press, Atlanta, USA, pp. 447–455
Sutinen, R. (1981): Causticizing plant and lime kiln computer control. Pulp Pap. Canada 82(8), 90–95
Uronen, P., Aurasmaa, H. (1979): Modelling and simulation of causticization plant and lime kiln. Pulp Pap. Canada 80(6), T162–T165
Uronen P., Leiviskä K., Aurasmaa H. (1975): Simulation study of the lime kiln, In: Hamza M. (ed.): Proceedings of Simulation '75, An International Symposium and Short Course, Zürich, p. 222–227
Uronen, P. and Leiviskä, K. (1989): New topics in lime kiln control. Pulp and Paper Canada 90(11), 113–117
Valiquette, J. (1999): Practical aspects of model-predictive control implementation on an industrial lime kiln. Tappi J. 82(5) 130–136
Zanovello, R., Budman, H. (1999): Model predictive control with soft constraints with application to lime kiln control. Computers and Chemical Engineering 23(6), 791–806
Yang, B., Yi, L. & Shouring, Q. (1997): A rule based cement kiln control system using neural networks. In: Proc. On IEEE International Conference on Intelligent Processing, Beijing, China Systems, vol. 1, pp. 493–497

Acknowledgements

The authors wish to thank the staff of the Wisaforest mill for their valuable support and assistance during this project.

Fuzzy Logic and SPC

Kimmo Latva-Käyrä
Tampere University of Technology, Automation Department, P.O. Box 692, FIN-33101 Tampere, Finland

1 Introduction

The use of statistical process control (SPC) procedures such as Shewhart charts has been one of the pillars of the quality revolution. The Shewhart charts display process data under test and provide statistical decision making tools to assess the process quality (Shewhart 1931). These simple charting methods have proven to be powerful tools for monitoring the 'health' of a process, for detecting special causes and events, and showing if the process in under control. In quality monitoring (Box and Luceño 1997) applications of process industry we must often use uncertain measurements and SPC is a good tool for handling them.

In paper mills there are lots of measurements and quality control based on those measurements. Operators can not observe all the measurement and also the data measured are more or less inaccurate. Our task was to help operators of the TMP plant to control the quality of the TMP pulp by calculating the degree of quality and displaying the most important quality and process variables in an easy form to the operators. This is carried out using a quality observer, which utilises cumulative sum charts and fuzzy logic tools. The system is tested under actual operation environment.

1.1 TMP-Process

Fig. 1 shows a simplified flowsheet of a typical TMP plant for producing pulp for mechanical printing papers. The wood logs are debarked and cut down to chips of a certain size. The chips are steamed in an atmospheric steaming vessel at 100°C, after which chips are washed in hot circulating water. A plug screw feeder feeds the warm and wet chips into a pressurised pre-heater with relatively low pressure and temperature. After the pre-heater, the chips pass a second plug screw feeder on their way into the first refiner, which operates at relatively high pressure and temperature (usually 300–500 kPa positive pressure and 143–158°C temperature). After this first-stage refining, the coarse pulp is blown to a steam separator from which it is fed to the second-stage refiner. From the second-stage refiner, which operates at about the same pressure and temperature as the first-stage refiner, the pulp is blown to a second steam separator. After passing the plug screw feeder at the bottom of the second steam separator, the pulp falls down into a pulper for removal of latency. Then follows screening and reject refining, and usually pulp is also bleached before storage and transport to the paper mill.

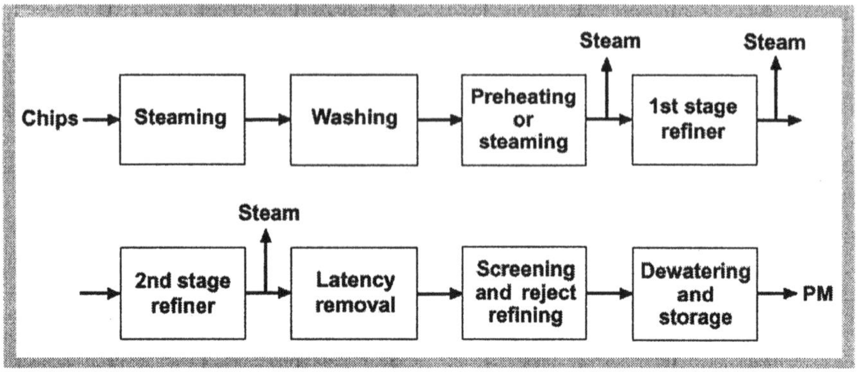

Fig. 1. Flowsheet of a TMP-plant.

1.2
Control Aspects for TMP-Refiner

Fig. 2 presents the control of the main line refiners (Sundholm 1999). The mostly used direct control variables for the TMP-refiner line are the refiner plate gap clearance and the flow rates of dilution water. (Strand 1995)These are related to two process factors: specific energy consumption (SEC, meaning refiner load divided by the production rate (Myllyneva et al. 1998) and refined pulp consistency. In the main line refining two refiners are cascaded. The power split between two refiner stages is also an important control variable because the refiner plate condition varies. At some plants, the plate gap clearance is not measured and plate vibration has to be used instead.

These basic refiner controls may be cascaded with the upper level pulp quality control (See also Chapter 11 of this book). Control of the refining is also a question about the selection of the most reliable measurements. This is the reason for the fact that the pulp quality control is done in most plants manually.

The most important on-line measurements of pulp quality are freeness (CSF), average fibre length (MFL) and content of shives. Fig 3 shows a comparison between different measurements from the primary refiner. The measurements have been made from a test run during two days period. The dependence has been strict among three process factors: refiner load, refined pulp consistency and plate gap. The dependence is not so clear when the comparison is made to the pulper freeness instead of the primary refiner load. The freeness and mean fibre length are dependent also on many other factors including raw material, the operation of the secondary refiner, mixing in pulper chest and sampling from pulper outlet flow. In this case no control operations were done on the secondary refiner. For the effective control of refiners, statistical averaging is needed when the control effect to the final pulp quality is monitored.

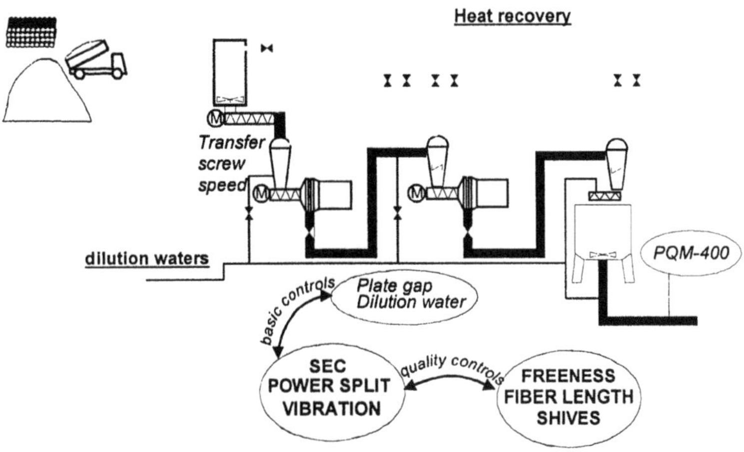

Fig. 2. Control of the main line refining. The process variables that should be stabilised are marked with normal font. The most important measured variables for the control of the main line refining are with upper case font. Main control variables are with italic font.

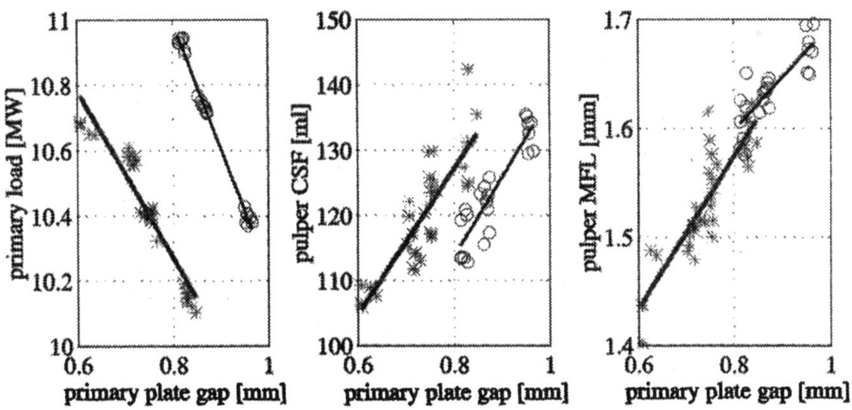

Fig. 3. The dependence between primary refiner control variables compared to the dependence on the pulp quality measurements in the pulper outlet flow. The different marks are for different refining consistency levels: '*' for low consistency and 'o' for high consistency. The effect of plate gap control operations can be seen directly on the refiner load, depending on the consistency level. The effect is not so clearly seen on the individual pulp quality values sampled after pulper and averaging is needed.

2 Development Environment

2.1 KCL-WEDGE

Quality observer is implemented to KCL-WEDGE process analysis system. It was chosen because it is an efficient platform to implement calculations and trend displays. KCL-WEDGE is a tool for production engineers and operators to monitor the process variations continuously, to analyse the variations systematically with proper mathematical tools and to compress the vast raw process data into information that can readily be utilised in operators' and engineers' decision making (KCL-WEDGE User Guide).

The distributed object extension of the WedgeLISP application language allows many KCL-WEDGE sessions to share an object and use it as if it was their own. Distributed objects of this kind are known as components. A method call of a component effectively passes data between two KCL-WEDGE sessions through arguments and a return value of a method enabling simple client/server programming. A server component can collect a pool of client components and make asynchronous calls back to the clients. The technique is known as Publish & Subscribe. Example of this can be seen at Fig. 4.

A monitoring KCL-WEDGE session serves as a publisher of warnings. The session executes an analysis whenever there is new data coming from the process being monitored. If the analysis detects an alarming situation, it publishes the warning to all the subscriber sessions currently in the subscriber pool. A session subscribes warning notifications by adding a client component to the subscriber pool. The session unsubscribes the notifications by removing the component from the pool.

2.2 Statistical Process Control

SPC is an optimisation philosophy concerned with continuous process improvements, using a collection of (statistical) tools for data and process analysis, making inferences about process behaviour and decision making. The goal of statistical process control is the elimination of variability in the process.

The Shewhart control charts, although they are very useful in detecting significant mean deviations or high data scatter, are not very effective in detecting specific faulty events (MacGregor 1988). In fact, they fail if very slow drifts or cyclic disturbances hidden in by the natural dispersion of the process have to be detected early. Using advanced control charts, based on further processing of measurement results, such as the cumulative sum (CUSUM) and the exponentially weighted moving average (EWMA) can handle these problems.

Shewhart chart techniques and other basic SPC tools are mainly meant for the applications in parts manufacturing industries. To apply them to the needs of the process industry requires further processing of these tools and the process knowledge. This is because the measurements are often uncertain with irregular sampling times and the measurement data is often sparse and correlated.

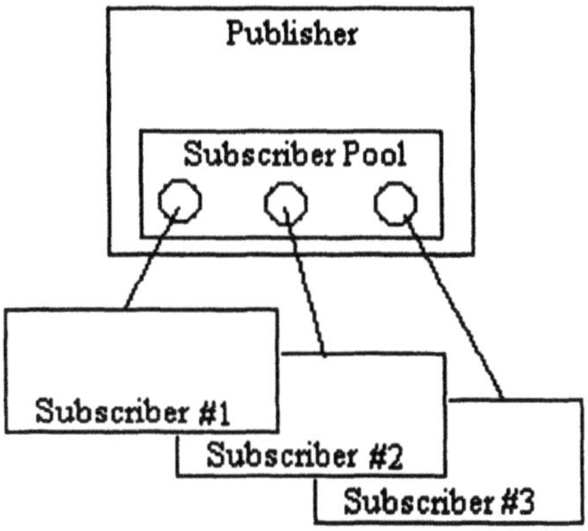

Fig. 4. Warning distribution using a subscriber pool

2.3
Exponentially Weighted Moving Average Charts

In the case of highly autocorrelated data the usual Shewhart-type control charts will be ineffective. Useful approach in many cases is to apply an EWMA control chart directly to the process data. EWMA is a general control scheme that is useful in many processes where the observations are positively autocorrelated and the mean does not drift too fast. The EWMA method can also be applied to the irregularly sampled data and sparse data can be handled with weight coefficients of EWMA.

EWMA is also a good alternative when small shifts and slow drifts are the main problems in the process. Normally EWMA is recursively calculated but in our case the data is irregularly sampled and sparse. So, we must use exponentially weighted moving window to calculate weight coefficients and then do the averaging. It is a bit slower than recursive calculation.

Weight coefficient of EWMA can be calculated as shown in equation 1.

$$m_i = Pe^{(t_n - t_{(n-i+1)})/\nu} \qquad (1)$$

where:

$$b = \frac{\ln(\frac{k}{P})}{win},$$

P is a maximum of weight coefficient,
k is a minimum of weight coefficient,
win is a window length.

Figure 5 shows how the weight coefficients of EWMA are formed.

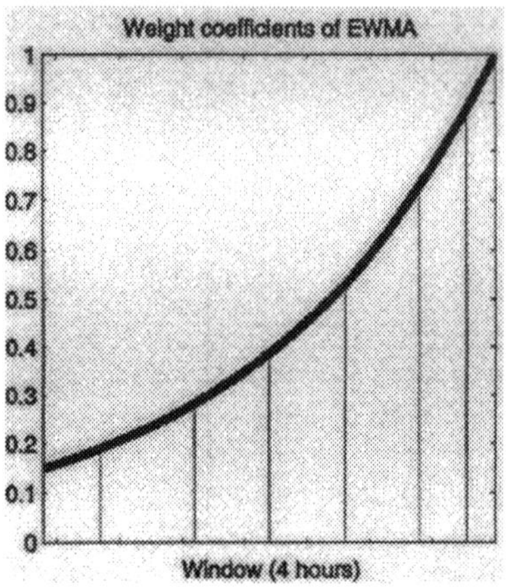

Fig. 5. EWMA weighting window when P is 1, k is 0.15 and *win* is 4 hours.

2.4 Relative Weights

The sum of the weight coefficients can be used to form the relative weight. This is a method for handling the sparse data based on statistical reliability i.e. dependency on the number of measurements in the processing window.

Fig. 6 shows an example of how the relative weight can be formed from the weight coefficients of EWMA. In this case the relative weight gets values between (-0.2, 0.2) but the values can depend on the process. Weighted average x_n can be calculated as shown in equation 2.

$$\hat{x}_n = \frac{\sum_{i=1}^{N_n} m_i x_{(n-i+1)}}{\sum_{i=1}^{N_n} m_i} \quad (2)$$

where:
x_n is a measurement at time t_n,
N_n is a number of measurements in *win* at time t_n.

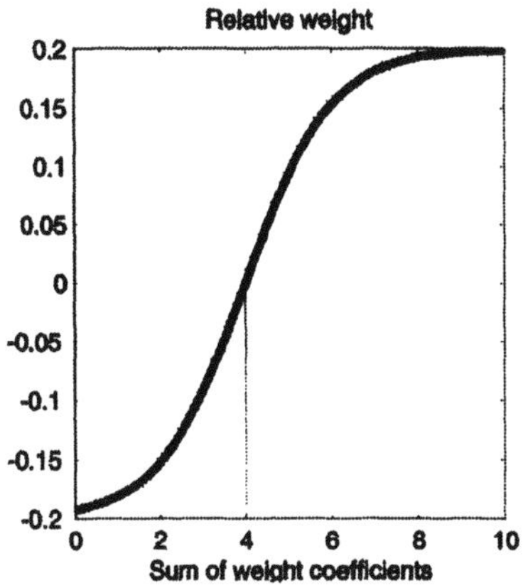

Fig. 6. Relative weight is based on sum of the weight coefficients.

2.5 Cumulative Sum Control Chart

The cumulative sum control chart is also a good alternative to the Shewhart control charts when we are interested in detecting small shifts (MacGregor 1988). CUSUM was chosen because it is excellent for detecting changes in means. In case of cumulative sum we assume that sampling is regular and calculate the cumulative sum normally. Cumulative sum can be calculated as shown in equation 3.

$$CUSUM_n = \sum_{i=1}^{N_n} (x_{(n-i+1)} - T) \qquad (3)$$

where:
T is a target value.

3 Fuzzy Extensions

3.1 Fuzzified Control Limits

Usually in the SPC the control limits are defined by standard deviation. Generally 3-sigma control limits are used. The EWMA control limits depend on the sample size (n) and the factor λ. Control limits for EWMA can be calculated as shown in equation 4.

$$CL = \overline{X} \pm A^{*} \sigma_x \qquad (4)$$

where:

$\overline{X}$ is the target value,
A^* is the control limit coefficient (deVor 1992).
σ_x is the standard deviation of x.

Fuzzy control limits for the EWMA control chart can be set up so that control limits given in equation 4 have a membership value between (0,1) depending on the chosen membership functions and how strict the limits should be. Figure 7 (flow diagram) shows that the sigmoid membership function was chosen for EWMA and CUSUM control limits (Latva-Käyrä 1998).

V-mask is a tool that is used in cumulative sum to detect the process variations. The detection process consists of placing the V-mask on the cumulative-sum control chart with the point 0 on the last value of the sum. Detection is based on the slope of the cumulative sum. If all the previous cumulative sums lie within the two arms of the V-mask, the process is in statistical control. Fuzzification is also suitable for the V-mask as shown in Fig. 8.

Quality observer uses these newer and more efficient SPC methods as exponentially weighted moving average (EWMA) and cumulative sum (CUSUM). Large database of measurements was used to determine the control limits and the parameters of the exponentially weighted moving average and cumulative sum. The database was also useful when the combined warning limit was defined.

3.2 Fuzzified Warning Limit

We have two different quality quantities, EWMA and CUSUM, and we can study them separately but we also want to look the fuzzy combined value. This combined warning limit is formed into EWMA-CUSUM co-ordinate system. It can be a circle or some kind curve. Fig. 9 shows the different possible warning limits in EWMA-CUSUM co-ordinates system.

Warning limit is formed from the fuzzified results of EWMA, CUSUM, the relative weight and the direction of the last measurement. Warning limit was defined by using simulation and large database of measurements.

3.3 Degree of Warning

Degree of warning tells the quality of the TMP pulp. The degree gets values between 0 and 1. Value 1 means loss in quality of the pulp. Fig. 7 shows flow diagram of how the degree of warning is calculated. Firstly data is collected, 4 hours data history is used, then SPC-methods, EWMA and CUSUM are calculated, then fuzzified and combined with each others and the fuzzified results from the relative weight and the direction of the last measurement. Combined warning limit is a monotonic curve in EWMA-CUSUM co-ordinates system.

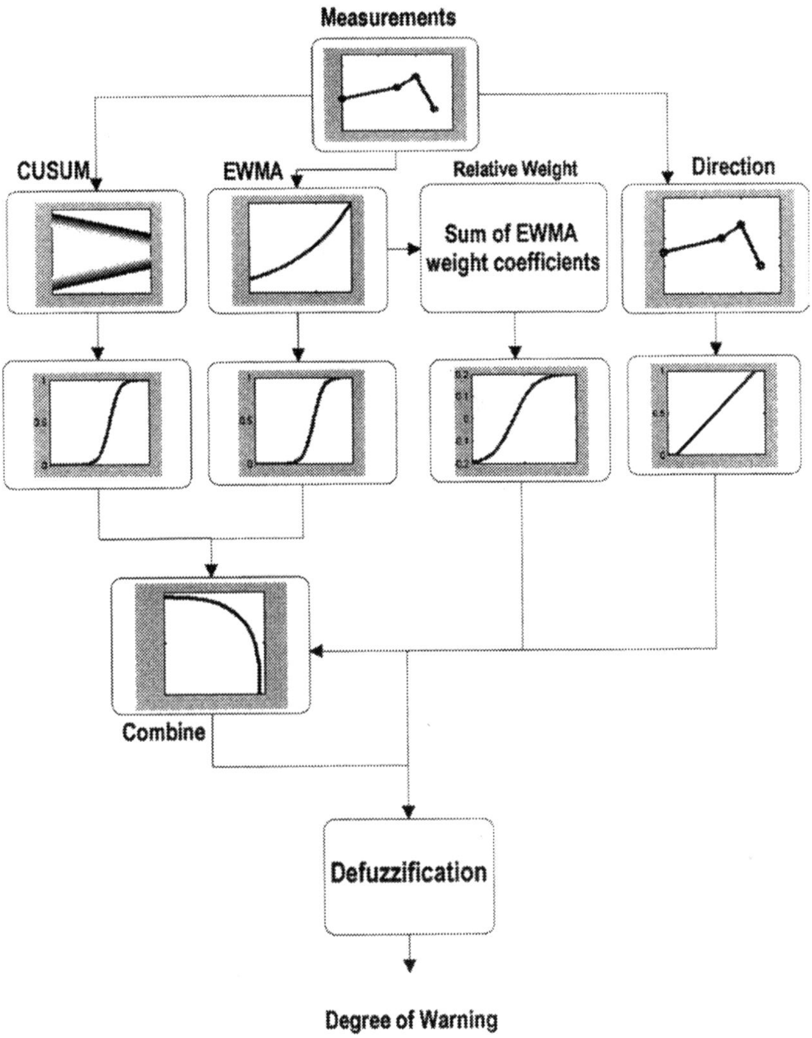

Fig. 7. Flow diagram of calculation of the degree of warning that tells to operator when the quality of the pulp is off the allowed limits.

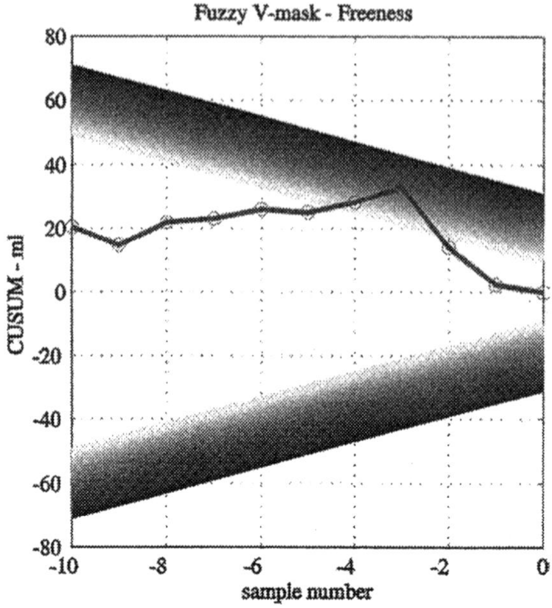

Fig. 8. Fuzzified V-mask used with cumulative sum.

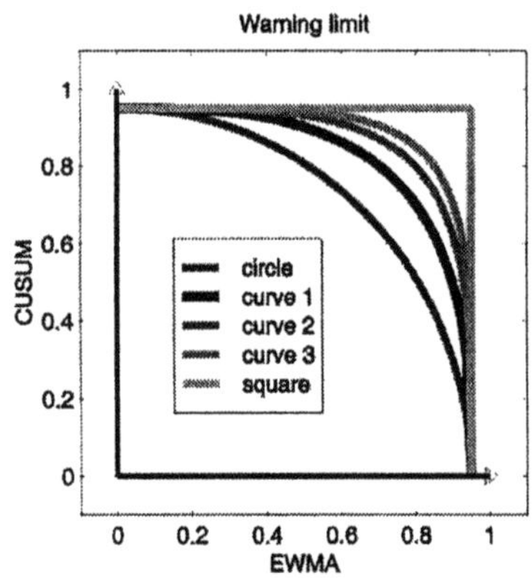

Fig. 9. Different possible warning limits.

4 Quality Observer

Quality observer is a piece of software and it was developed for UPM-Kymmene Jämsänkoski TMP2-plant and its main purpose is to monitor mechanical pulp quality in TMP plant. This task is difficult for the operators because of the many parallel production lines in the plant. Also the data measured from these lines are more or less inaccurate. Two on-line quality measurements, freeness and fibre length, are monitored.

The application is made for operators to help them to control the quality of the TMP pulp. Graphical user interface (GUI) was designed to be easy to use and to be very informative. Fig. 10 shows an example of the quality observer GUI.

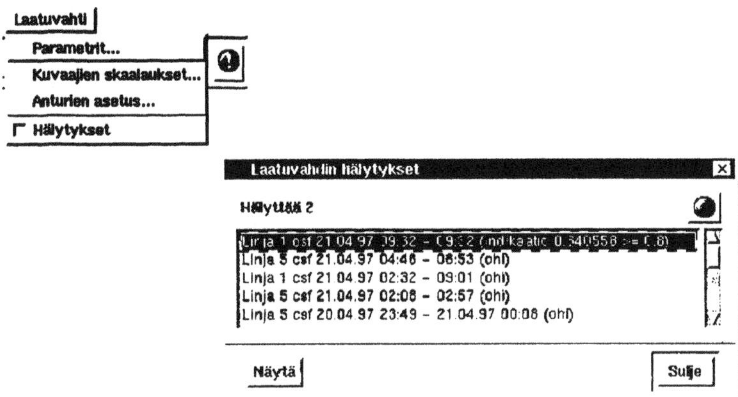

Fig. 10. Example of the user interface of the quality observer. From the menu in the upper left corner the operator can choose parameters, scaling of diagrams and sensors, or look at the warnings. Existing warnings are shown in the right-hand side of the display.

Every time a measurement is made, the quality observer is called. When it detects that either variable has exceeded the allowed quality limits, it tells the operator the information of which variable caused the quality variation and in which line it happened. There is also a possibility to look the trend display. Trend display contains data from all associated process variables (freeness, fibre length, SEC (specific energy consumption), power split and plate vibrations) for several hours before the warning occurred. In practice this kind of display is very helpful to localise the causes of the detected variation. Also the possible earlier warning events can be seen in the trend display. Fig. 11 shows an example of a trend display.

5 Results

Table 1 shows a comparison between the different warning limits. Curves are a monotonic in both directions. Testing of different warning limits was made by simulation using large database. The detected warning was right if there were enough alarming measurements after the detected variation.

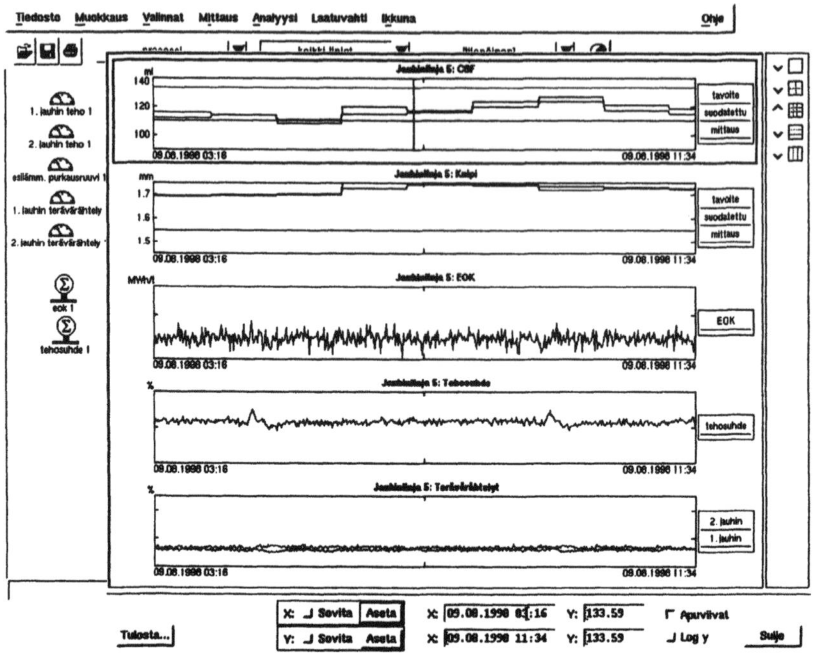

Fig. 11. KCL-WEDGE screen and a trend display with an example of the quality observer warnings (grey box in the CSF trend). Trends include from top freeness, fibre length, SEC, power split and plate vibrations. Freeness has exceeded the allowed quality limits and quality observer has published the warning.

Table 1. Comparison between the different warning limits and scaling methods.

Warning limit	Scaling method	Right (%)
Circle	1	87,68
Circle	2	87,17
Curve 1	1	88,33
Curve 1	2	87,10
Curve 2	1	88,39
Curve 2	2	87,49

Fig. 12 shows weighting of SPC procedures in EWMA vs. CUSUM plane. The curve 1 was chosen because it detected more alarming situations than curve 2 although the chance for being right is slightly smaller. The relative weight is 0 in the Fig. 12. The relative weight and the direction of the last measurement also affect the final degree of warning as seen in Fig. 7.

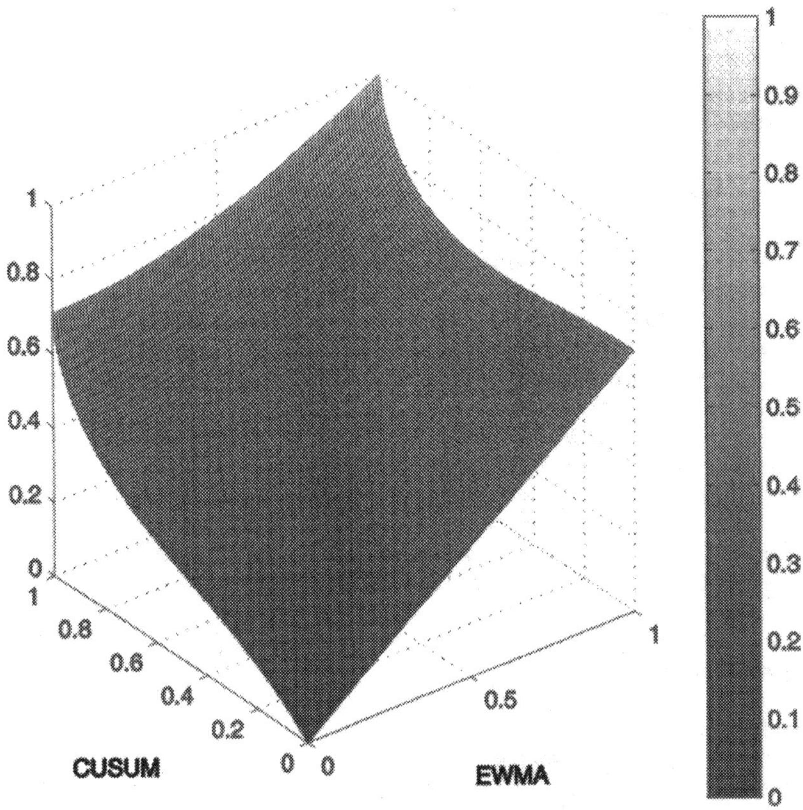

Fig. 12. Warning limit with curve 1 and scaling method 1.

6 Conclusions

Statistical process control is an important tool for detecting quality variations in the process. Good results can be achieved by using advanced SPC methods like EWMA and CUSUM with fuzzy control limits and their fuzzy combination. The quality observer uses those methods to monitor the quality of the TMP pulp. It monitors two on-line quality measurements, freeness and fibre length and when either has exceeded the allowed limits the operator is informed.

The quality observer was developed for UPM-Kymmene Jämsänkoski TMP2-plant. The observer was implemented in their KLC-WEDGE process analysis system. Several years database was used to test the methods and tune the observer. From the basis of this data it was verified that the correlation between the on-line analyser and the laboratory analysis has been good for both the freeness and the fibre length during recent years.

The future plans include the multivariate analysis based on combined data of all the refiner lines, with target to divide the causes for quality variation to common disturbances for all refiner lines and to disturbances unique for one refiner line. Also the use of dynamic calibration is in our plans.

References

Shewhart, W.A. (1931): Economic Control of Quality of Manufactured Product, D. Van Nostrand, New York, USA

Box, George, Luceño, Alberto. (1997): Statistical control by monitoring and feedback adjustment. A Wiley-Interscience Publication. 327 p

Papermaking Science and Technology - Mechanical Pulping, ed. Jan Sundholm (1999), Fapet Oy, Helsinki, Finland

Strand, B.C. (1995): The effect of refiner variation on pulp quality. In: International Mechanical Pulping Conference Proceedings, CPPA, Ottawa, pp. 125 - 130

Myllyneva, J., Kortelainen, J., Latva-Käyrä, K., Nystedt, H. and Leiviskä, K. (1998): Hiertämön laatusäädöt [HILA] (Quality Control of the TMP-Plant). Final Report. University of Oulu, Control Engineering Laboratory, Oulu, 48 p. (in Finnish)

Montgomery, D. C., (1991): Introduction to Statistical Quality Control, 2nd ed. John Wiley & Sons, New York, USA.

MacGregor, J. F. (1988): On-line Statistical Process Control, Chemical Engineering Progress 84(10), pp. 21-31

DeVor, R., Chang, T. and Sutherland, J. (1996): Statistical Quality Design and Control. Macmillan Publishing Company

Latva-Käyrä, K. (1998): TMP pulp quality monitoring, TTKK/MIT, Tampere, 83 p. (in Finnish)

KCL-WEDGE, User's Guide. KCL Development Oy. 121 p.

Estimation of Target Values for Grade Changes in Paper Machine

Pauli Viljamaa[1], Heikki Peltonen[2], Heikki N. Koivo[3]
[1]Tampere University of Technology, Automation and Control Institute, P.O. Box 692, FIN-33101 Tampere, Finland
[2]Process Research, Laserkatu 6, FIN-53850 Lappeenranta, Finland,
[3]Helsinki University of Technology, Control Engineering Laboratory, PL 5400, FIN-02015 TKK, Finland

1 Introduction

Paper making is a continuous process, where up to 10 m wide paper web is produced and fed through a paper machine (Fig. 1) at the speed up to 1800 m/min. The paper machine may produce paper 300,000 tons per year stopping only for short annual maintenance breaks. These numbers are approximate but give an idea how large-scale system a paper machine is.

A paper machine may have dozens different paper grades in its production repertory. A grade change on a paper machine means changing the current paper grade to a new one. Grade change is made by changing process variables such as speed, stock flow and steam pressures at the same time to provide a fast transition in quality variables as basis weight and moisture (Menani et al. 1998).

Storing paper should be avoided in paper mills and all production tries to run only for one order. Therefore the size of the batches in the orders effects the grade scheme and the amount of grade changes. Nowadays a paper machine may have at least a grade change per day. Usually the grade change takes time from few minutes to a half an hour to be completed. The range is wide, because the time depends on the size of the change.

The paper made during the grade change is often out of quality tolerances of both grades. This off-quality product is handled like a broke and this increases production costs (Ranta et al. 1992). For example on LWC paper machine due to off-quality production during grade changes, the decrease of sales income can be worth 4,000,000 USD/year (Viitamäki 1997). That is why the grade change is wanted to be done as fast as possible.

For the sake of the complexity of the process the grade change is a very demanding operation. To help operators' work quality systems often have special automatic grade change programs. Typically grade changes are done with quality controllers in open loop. The grade change program gives ramps for the set points of low-level controllers to the new target values. Timing of the ramps and the target values are critical for successfulness of the grade change (Fig. 2). If the given target value is not correct, it may cause quality variations of the product after the machine direction controls are switched on. This study concentrates on computing the target values for the set point of the low level controller.

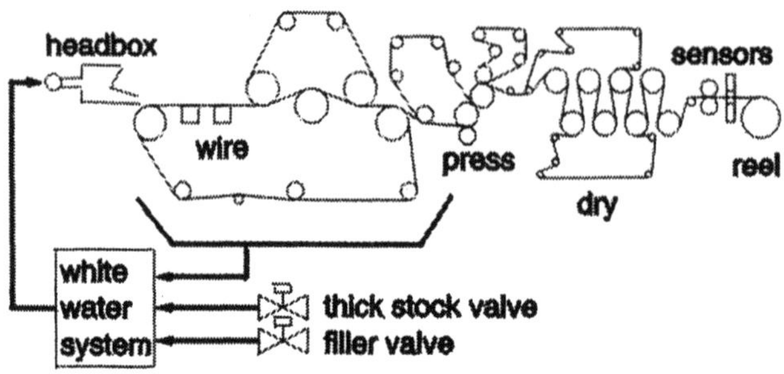

Fig. 1. A fine paper machine (Makkonen et al. 1994).

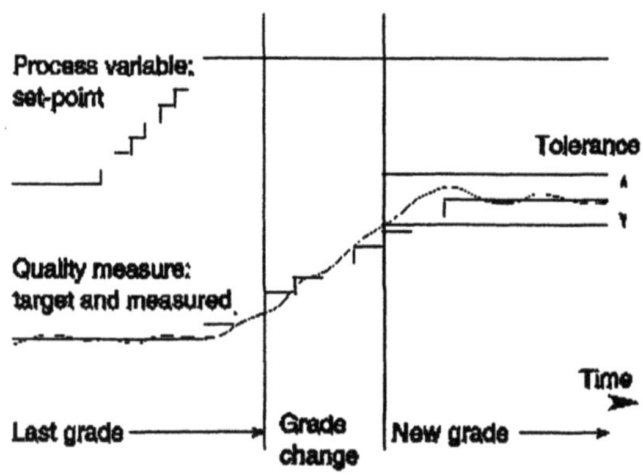

Fig. 2. Change of quality measure in a grade change.

This study proposes a method to utilize a fuzzy system to compute target values during grade changes. It shows an industrial example of this (Viljamaa et al. 2000). As far as the authors know this is the first application of fuzzy computing to automatic paper machine grade change.

The paper is organised as follows. The problem is introduced in Section 2. The solution for the problem is proposed in Section 3. The significance of the expertise is discussed in Section 4. The industrial example is shown in Section 5 and conclusions are given in Section 6.

2 Problem Statement

The main goal of this study is to develop a non-linear model to estimate the new target value for the set point of the low level controller. The model should be capable of estimating the new set point for the low level controller when the current set point and the quality measures for the current and new grades are known. The estimate is computed in one shot when the grade change is started.

Practical aspects set some limits for the modelling method. The main problem is that each grade change is unique. Although two grade changes have similar changes, in practice the state of the paper making process differs at each time. The process is non-linear and time variant. That is why the modelling data of the process cannot be from a long time period and thus, amount of data needed for the modelling should be small.

For maintenance reasons it is better that the operators can understand action of the model and effect of the parameters (Sanders 1998). Additionally, complexity of the model must be taken into account in considering possibilities to update the model. Thus, a pure black-box model, e.g. a neural network, is not a very suitable alternative.

3 Problem Solution

Fuzzy systems provide a possibility to transfer heuristic knowledge into a conventional mathematical representation (Viljamaa 2000), which furthermore can be tuned using input-output data (Lindskog 1996). Additionally, they are transparent to interpretation and analysis (Babuska 1998; Viljamaa 2000). Thus, fuzzy models are selected as a method to solve the problem.

Standard structure of a fuzzy system (Jager 1995) with product-sum composition, triangle shaped input membership functions, singleton output fuzzy sets and centre-of-gravity defuzzification are utilised because of many reasons. Product-sum composition performs more smoothly than min-max composition (Viljamaa and Koivo 1998). Triangular membership functions provide universal approximation property (Zeng and Singh 1996) and true transparency. The latter means that the output of the system is the same that can be assumed based on the rules (Viljamaa 2000). True transparency cannot be obtained with input membership functions which have unbounded support. Singleton output fuzzy sets mean that the system can be seen as a special case of Takagi-Sugeno model with constant output functions (Babuska 1998).

The fuzzy model is developed with the operators of the paper mill. They can select the important variables as inputs of the model. They can express the relations between the inputs and the outputs in a verbal form. It makes the basis for the model building.

Fuzzy modelling scheme provides a convenient way to include the knowledge the operators have into the model. It is possible to build the whole model with them and test it thereafter. When input-output data is available, the model can be tested immediately. If the performance is unacceptable for a data sample, rules affecting that can be found, and often the performance can be improved adjusting the rules. Because the rules are local the adjustment does not usually degrade the performance elsewhere (Viljamaa 2000).

An additional difference of the fuzzy system and the linear model is that the parameters of the fuzzy model can be straightforwardly optimised in such a way that a priori knowledge is included into the model. This is achieved by fixing the consequences of certain rules before the optimisation. Similar fixing can be done, e.g. in the case that a rule is located in a region where no information is available in the optimisation data.

Also the varying interval of the output can be directly seen from the parameters. This is useful in considering the validity and safety of the model. Additional limit checking is unnecessary after the consequence parameters are validated. For example, even simple linear models need this checking in order to avoid possible large output values.

4
Importance of Expertise

Although fuzzy computing has an easy and natural way to represent heuristic knowledge in the form of linguistic if-then rules, a good implementation of the reasoning needs expertise that personnel at industry do not necessarily have. The experts of the applications area need help of experts on the field of fuzzy computing. If the process experts try to solve the problem by themselves, it is usually too demanding task to do within such a tight time schedule the industry often have. The development project goes more smoothly when the expertise in different areas is combined with.

The need of fuzzy expertise lies in the fact that there are so many alternatives in the field of fuzzy computing, e.g. see (Driankov et al. 1996). What kind of operators are needed, is it worth to build the rule base utilizing all fuzzy operators like and, or, and not? How to implement those operators, what are good solutions for T-norm and S-norm? For example, T-norm and S-norm have several possible implementations presented, e.g., in (Gupta and Qi 1991). How does a selection affect properties of the system? One such comparison is presented, e.g., in (Viljamaa and Koivo 1998), but such results are not well known currently. These questions are discussed in (Viljamaa 2000).

The shape of the membership functions has an effect on the smoothness of the output surface of the fuzzy system. But what kind of smoothness do we need? How to place the membership functions? And how many membership functions should we have? Should we select defuzzification method A or B? All these are common questions in practice, but no good cookbook is available where those problems would be addressed clearly.

An important question is, how to capture the knowledge the process experts have into the form of fuzzy rules. Although many clustering methods (Babuska 1998) and optimisation algorithms (Lindskog 1996) are developed for the purpose to extract the rule base from data, they can not replace human knowledge and creativeness. Usually the operators of the mill have the best knowledge of the process and therefore the fuzzy system can be generated very successfully by interviewing them. The preparation for the interviews needs basic knowledge of the possibilities the fuzzy computing provides. And usually the best results are obtained when those methods are combined in a reasonable way.

More research is needed from the practical point of view. Before such a results are available the process experts will prefer the help from experts of fuzzy computing. That was also observed in the application presented in this paper.

5 Industrial Example

This section describes a case study, which has been carried out in an industrial paper mill. The problem concerns adjusting the set points during grade changes. The aim of the project was to build a system, which computes new set points for the controllers.

The purpose of this example is to illustrate an iterative design process of a fuzzy system. It shows which kind of methods was utilised with personnel of the paper mill. The section makes some remarks on the principle differences between linear and fuzzy models clearly seen in practice.

This study reports only the initial building stage of the system and does not concern the following actions to implement the system. However, the system developed in the project has also been taken to use.

The development was performed in a meeting with experts from the mill. They selected a huge list of variables correlating with the output. The list was first considered more critically. The most significant variables were selected and their priorities were given. The group decided to estimate the change in the set point and two of the most significant variables, and, were selected as inputs for the model.

The building of the linguistic rule base was started by defining fuzzy labels for the inputs. Input x_1 had initially three fuzzy labels, negative (N), zero (Z) and positive (P) as a default. But the number of them was increased to five in order to add some capability to create non-linear behaviour with the fuzzy rules. Therefore, negative big (NB), negative small (NS), zero, positive small (PS) and positive big (PB) were adopted. Three fuzzy labels were defined for input x_2, small, medium and big.

The antecedents for the rule base were generated systematically resulting 15 propositions like "x_1 is small and x_2 is small". The rule base had five rules where "x_2 is medium", which was considered as a normal situation. The consequences were generated for those rules according the input x_1. Additionally other rules with the proposition "x_1 is zero" were set to zero. Thereafter the rules with the propositions "x_2 is small", and "x_2 is big" were adjusted according to the neighbouring "normal" rules. In Table 1(a) the symbols, e.g., NS- and NS+, mean that the corresponding rule is adjusted toward or away zero from the original location, e.g., NS.

Then numerical values were given to the rule base according to the knowledge of the experts. The numerical values presented in Table 1(b) were obtained. The numerical values mean the core points of the triangular membership functions (the two left-hand columns) and the places of the output singletons (the right-hand column).

The system was tested with the collected data and it was detected that at one grade change the system performs very poorly (dashed line in Fig. 3). Otherwise the accuracy was surprisingly good. The dominant rule for the poor estimate was found and improved. The consequence of the last rule in Table 1 was changed to the form of y_1 = 0.6. The change improved the performance at four additional grade changes (solid line in Fig. 3).

Table 1. The rule base of the first try.

x_1	X_2	y_1	x_1	x_2	Y_1
NB	S	NB+	-0.5	0.1	-1
NB	M	NB	-0.5	0.3	-0.8
NB	B	NB-	-0.5	0.6	-0.6
NS	S	NS+	-0.3	0.1	-0.6
NS	M	NS	-0.3	0.3	-0.4
NS	B	NS-	-0.3	0.6	-0.3
Z	S	Z	0	0.1	0
Z	M	Z	0	0.3	0
Z	B	Z	0	0.6	0
PS	S	PS-	0.3	0.1	0.3
PS	M	PS	0.3	0.3	0.4
PS	B	PS+	0.3	0.6	0.6
PB	S	PB-	0.5	0.1	0.6
PB	M	PB	0.5	0.3	0.8
PB	B	PB+	0.5	0.6	1

(a) Linquistic. (b) Numeric

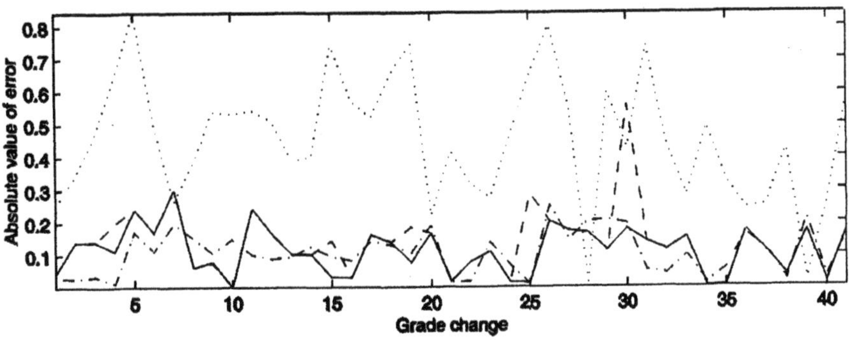

Fig. 3. The absolute errors of the fuzzy model from Table 1(b) (dashed line), the improved fuzzy system (solid line) and a linear model with a bias (dot-dashed line) with respect to grade changes. The first 30 were used for optimisation of the linear model. The actual change in the output (absolute value) is shown by the dotted line. The horizontal dotted line shows the acceptable error limit.

After the fine adjustment the estimation error of the fuzzy system (solid line) is four times outside of the tolerance limit (0.2) as a linear system with a bias (dot-dashed line) does as well (Fig. 3). The parameters of the linear system were obtained by the least squares estimation with respect to the first 30 grade changes.

Some iterations were made in order to test if additional inputs could explain the estimation error. First, one input with three fuzzy sets was added to the rule base resulting in 45 rules. The operators were capable of expanding the rule base in quite a short time with the same principles as the rules in Table 1(a). The mean value of the absolute error was decreased but the maximum of the absolute error not.

The second additional input with two fuzzy sets increased the number of rules to 90. Furthermore, it was not possible to reduce the maximum of the absolute error. Both of the rule bases were refused because of the complicated structure and negligible improvements in the accuracy.

Finally, a pure optimised fuzzy system was created with two inputs (three fuzzy sets for each). At the same time, the first input was changed. The input membership functions were determined by the operators, but the consequences (Fig. 4) were optimised by a least squares algorithm. This approach is explained, e.g., in (Viljamaa 1996).

The resulting rule base is shown in Table 2 and the output surface in Fig. 5. The estimation error was compared with respective a linear model with a bias (Fig. 6). The mean value of the absolute errors and the maximum errors were remarkably smaller (15 % and 25 % with validation data) for the fuzzy model than for the linear model.

An additional remark about the differences of the fuzzy system and the linear model is that the parameters of the fuzzy model can be straightforwardly optimised in such a way that the system does not suggest a change in the output if the operation conditions do not change. This is achieved by fixing the consequences of the rules with propositions "x_3 is zero" to zero before the optimisation. Similar fixing can be done, e.g. in a case that a rule is located in a region where no information is available in the optimisation data. Also the varying interval of the output can be directly seen from the parameters. This is useful in considering the validity and safety of the model.

Table 2. Numerical rule base of the final try.

x_3	x_2	Y_1
-0.8	0.1	-0.37
-0.8	0.3	-0.67
-0.8	0.6	-0.92
0	0.1	0
0	0.3	0
0	0.6	0
0.8	0.1	0.74
0.8	0.3	1.06
0.8	0.6	0.74

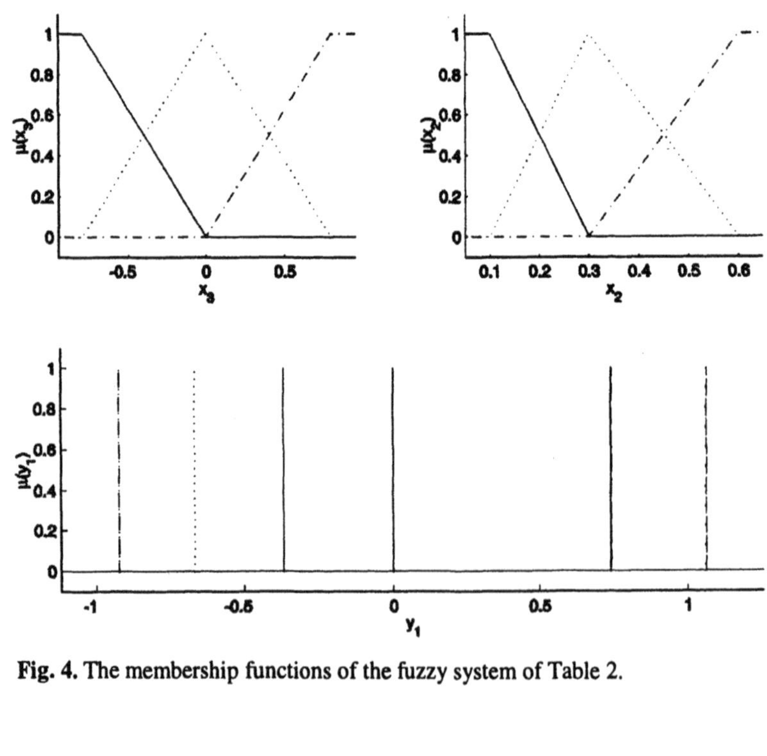

Fig. 4. The membership functions of the fuzzy system of Table 2.

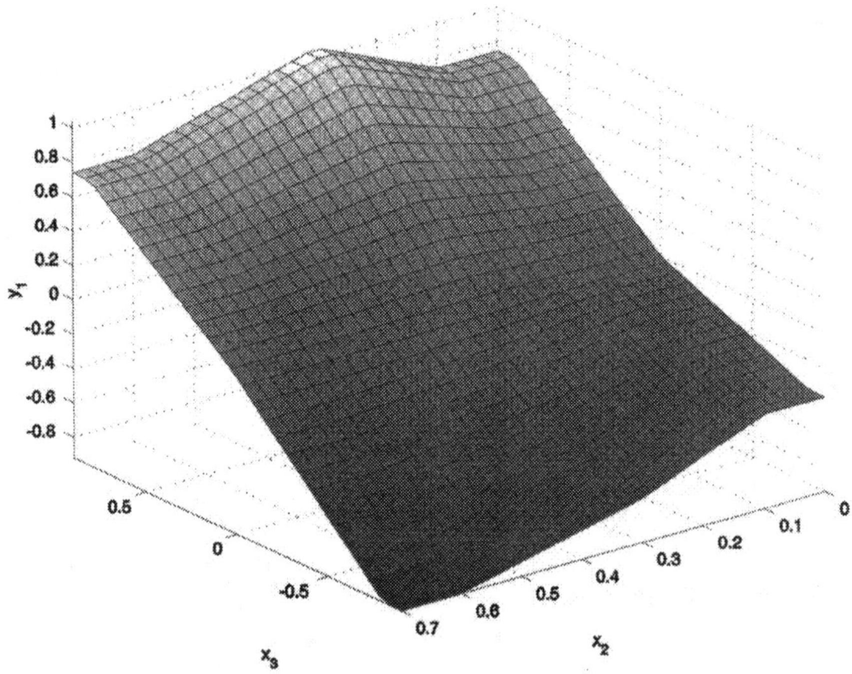

Fig. 5. The output surface of the fuzzy system of Table 2.

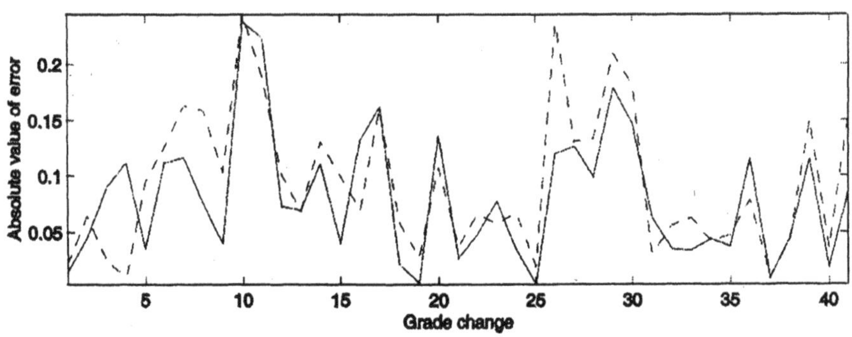

Fig. 6. The absolute errors of the fuzzy model from Table 2 (solid line) and a linear model with a bias (dashed line) with respect to grade changes. The first 30 were used for optimisation. The horizontal dotted line shows the acceptable error limit.

The results were so good that both the linear model and the fuzzy system were implemented in the paper mill and action of them is investigated in real environment. The operation of the models has been evaluated over a long time period. In each examination the fuzzy model has been more accurate and it has been selected for production.

6 Conclusions

An application of fuzzy computing is presented. Fuzzy modelling scheme is adopted to estimate the new target value for the set point of the low level controller during a paper machine grade change. The estimate is computed at the beginning of the change. This is the first time when fuzzy computing has been used and implemented in an industrial scale paper machine grade change. Operators are able to build the system in a reasonable short time. The parameters of the model can be tuned manually or estimated based on input-output data. The fuzzy model obtained by optimising techniques is still transparent to be interpreted and analysed by the operators.

Acknowledgements

The author (P.V.) wishes to thank the Graduate School in Electronics, Telecommunications and Automation (GETA), and the Academy of Finland for their support.

References

Babuska, R. (1998): Fuzzy modeling for control. Kluwer Academic Publisher, Boston

Driankov, D., Hellendoorn, H., Reinfrank, M. (1996): An introduction to fuzzy control. Springer-Verlag, Berlin Heidelberg New York, 2nd revised edition

Gupta, M.M., Qi, J. (1991): Design of fuzzy logic controllers based on generalized T-operators. Fuzzy Sets and Systems 40, 473–489

Jager, R. (1995): fuzzy logic in control. PhD thesis, Technische Universiteit Delft

Lindskog, P. (1996): Methods, algorithms and tools for system identification based on prior knowledge. PhD thesis, Linköping University, Linköping, Sweden.

Makkonen, A., Rantanen R., Kaukovirta, A., Koivisto, H., Lieslehto, J., Jussila, T., Koivo, H.N., Huhtelin, T. (1994): The paper machine benchmark for control systems. In: Control Systems 94 Conference. STFI, Stockholm, Sweden, pp. 219–225

Menani, S., Koivo, H.N. Huhtelin, T., Kuusisto R. (1998): Dynamic modelling of paper machine from grade change data. In: Control Systems '98 Proceedings. Finnish Society of Automation, Porvoo, Finland, pp. 79–86

Ranta, J., Ollus, M., Leppänen, A. (1992): Information technology and structural change in the paper and pulp industry. Computers in Industry 20, 255–269

Sanders, F.F. (1998): Advanced control success depends on operator confidence. InTech 45(9), 48–51.

Viitamäki, P. (1997): Modeling of paper machine grade changes: A multiple model approach. In: AIChE 1997 Annual Meeting, Los Angeles, USA

Viljamaa, P. (1996): Optimization of rule base for fuzzy logic system. In: Proceedings of TOOLMET'96 Symposium – Tool Environments and Development Methods for Intelligent Systems. University of Oulu, Oulu, Finland, pp. 159–164

Viljamaa, P. (2000): Fuzzy gain scheduling and tuning of multivariable fuzzy control–Methods of fuzzy computing in control systems. PhD. Thesis, Tampere University of Technology, Tampere, Finland

Viljamaa, P. Koivo, H.N. (1998): Comparison of minimum and product operators in fuzzy systems with more than two inputs. In: IASTED Int. Conf. on Control and Applications CA'98, Honolulu, HI, USA, pp. 69–73

Viljamaa, P., Peltonen H., Koivo, H.N. (2000): Target value computing in paper machine grade changes by fuzzy system. In: The 9th IEEE International Conference on Fuzzy Systems FUZZ-IEEE 2000, San Antonio, TX, USA, pp. 105–109

Zeng, X.-J., Singh, M.G. (1996): Approximation accuracy analysis of fuzzy systems as function approximators. IEEE Transactions on Fuzzy Systems 4(1), 44–63

Index

List of Contributors

Ahava, O.
UPM-Kymmene
P.O. Box 42, FIN-68601 Pietarsaari, Finland

Hyötyniemi, H.
Professor, Control Engineering Laboratory, Helsinki University of Technology
P.O. Box 5400, FIN-02015 HUT, Finland
heikki.hyotyniemi@hut.fi

Isokangas, A.
Research Assistant, University of Oulu, Control Engineering Laboratory
P.O.Box 4300, FIN-90014 Oulun yliopisto, Oulu, Finland
Ari.Isokangas@oulu.fi

Joensuu, I.
Product Manager, ABB Industry Oy, Systems Group
Tyrnäväntie 4, FIN-90100 Oulu, Finland
ismo.joensuu@fidri.mail.abb.com

Juuso, E.
Head Assistant, University of Oulu, Control Engineering Laboratory
P.O.Box 4300, FIN-90014 Oulun yliopisto, Oulu, Finland
Esko.Juuso@oulu.fi

Järvensivu, M.
Senior Consultant, Pronyx Control Software
Veneentekijäntie 4, FIN-00210 Helsinki, Finland
mika.jarvensivu@pronyx.fi

Karlsson, L.
Holmen Paper, Hallsta Paper Mill
S-76381 Hallstavik
lennart.karlsson@hallsta.holmen.modogroup.com

Koivo, H.
Professor, Helsinki University of Technology, Control Engineering Laboratory,
P.O.Box 5400, FIN-02015 HUT, Finland
Heikki.Koivo@hut.fi

Lassus, L.
Project Manager, Oy L M Ericsson Ab
FIN-02420 Jorvas, Finland

Latva-Käyrä, K.
Researcher, TUT/Measurement and Information Technology
P.O. Box 692, FIN-33101 Tampere, Finland
Kimmo.Latva-Kayra@mit.tut.fi

Leiviskä, K.
Professor, University of Oulu, Control Engineering Laboratory
P.O.Box 4300, FIN-90014 Oulun yliopisto, Oulu, Finland
Kauko.Leiviska@oulu.fi

Miettunen, J.
Automation Engineer, Outokumpu Mining Oy, Pyhäsalmi Mine
P.O. Box 51, 86801 Pyhäsalmi, Finland
jorma.miettunen@outokumpu.com

Myllymäki, P.
R & D Engineer, Ironmaking, Rautaruukki Steel
P.O. Box 93, FIN-92101 Raahe, Finland
pekka.myllymaki@rautaruukki.fi

Myllyneva, J.
Researcher, University of Oulu, Control Engineering Laboratory
P.O.Box 4300, FIN-90014 Oulun yliopisto, Oulu, Finland
Jaakko.Myllyneva@oulu.fi

Palovaara, T.
R & D Engineer, Steelmaking, Rautaruukki Steel
Primary Products, P.O. Box 93, FIN-92101 Raahe, Finland
tomi.palovaara@rautaruukki.fi

Peltonen, H.
Managing Director, Prosessitutkimus Oy
Laserkatu 6, FIN-53850 Lappeenranta, Finland
Heikki.Peltonen@kareltek.fi

Poutiainen, J.
Project Manager, R&D, Filtronic Comtek Finland
Takatie 6, FIN-90440 Kempele, Finland
jussi.poutiainen@lkproducts.fi

Rauma, T.
Group Manager, VTT Electronics/Intelligent Systems Group
P.O Box 1100, FIN-90571 Oulu, Finland
Tapio.Rauma@vtt.fi

Saxén, H.
Professor, Heat Engineering Laboratory, Åbo Akademi University
Biskopsgatan 8, FIN-20500, Finland
Henrik.Saxen@åbo.fi

Sipola, M.
R & D Engineer, Steelmaking, Rautaruukki Steel
Primary Products, P.O. Box 93, FIN-92101 Raahe, Finland
matti.sipola@rautaruukki.fi

Viljamaa, P.
Research Engineeer,TUT/Automation and Control Institute
P.O.Box 692, FIN-33101 Tampere, Finland
pauli@ae.tut.fi

Ylinen, R.
Professor, University of Oulu, Systems Engineering Laboratory
P.O.Box 4300, FIN-90014 Oulun yliopisto, Oulu, Finland
Raimo.Ylinen@oulu.fi

Yliniemi, L.
Laboratory Manager, University of Oulu, Control Engineering Laboratory
P.O.Box 4300, FIN-90014 Oulun yliopisto, Oulu, Finland
Leena.Yliniemi@oulu.fi

www.ingramcontent.com/pod-product-compliance
Ingram Content Group UK Ltd.
Pitfield, Milton Keynes, MK11 3LW, UK
UKHW020140290926
15471UKWH00003B/604
9783662003558